Young Folks

a hundred thousand children testify is expected
uch solicitude as rarely agitates the juvenile
It is published monthly, and contains the most
ful variety of TALES, SKETCHES, POETRY, GAMES,
&c., etc., and is embellished with a multitude of
FUL ILLUSTRATIONS by the BEST ARTISTS. Terms,
ar; a large discount to clubs. A specimen num-
be sent on receipt of 20 cents by the publishers.
TICKNOR & FIELDS, Boston, Mass.

O'DOR! O'DOR!
DR. BRIGGS'S
GOLDEN O'DOR

forces a Beautiful Set of Whiskers or Moustache
Smoothest face from five to eight weeks, within or injury to the skin, or Hair on Bald Heads
4 weeks (AND NO HUMBUG). I receive RECOM-
MENDATIONS most every day from persons that have
tried and found it genuine. Testimonials of thou-
I will send Dr. Briggs's Golden O'Dor by
mail and postpaid for $1 25. ☞ WARRANTED.
l orders to
DR. C. BRIGGS, Drawer 6308, Chicago, Ill.
x30

the New Army Corps Badges Ready.

T. HAYWARD,
208 Broadway, New York.

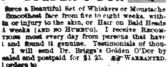

re a NEW PRESIDENT LINCOLN MEDAL, with
Likeness on one side, and the Date of his Birth,
nd second Inauguration, and of his Assassination.
le, 25 cents; $15 per hundred. Agents wanted
where.
the NEW RICHMOND MEDAL. On one side
Capture of Richmond, April 3, 1865, the opposite
f this Medal is a True Likeness of Gen. Grant.
Send a sample of this Medal without the Top Bar
cents, and with the Top Bar, made of Pure Coin
, with your Name, Regiment and Company hand-
ly engraved thereon, for $3.
I have all the NEW CORPS BADGES of the
ready. I will send a sample, with your Name,
ment and Company handsomely engraved thereon,
receipt of $1 50; and for $3, or $5, I will send
nine 16 karat Gold Enameled Corps Ring for either
or Division.
ges of every description made to order.
ant an Agent in every Regiment, Hospital and De-
ment of the Army and Navy, to whom especial
vements are offered. Send for wholesale Illustrated
lar.

B. T. HAYWARD,
Manufacturing Jeweller, 208 Broadway, N. Y.

THE BOWEN MICROSCOPE,

lfying 500 TIMES, mailed to any address for 50 cts.
of different powers for $1. Address
F. B. BOWEN, Box 220, Boston, Mass.

STEREOSCOPIC VIEWS
OF THE WAR!

ITCH (**WHEATON'S**) **ITCH**
SCRATCH. OINTMENT SCRATCH.

Will cure the Itch in 48 hours—also cures Salt Rheum,
Ulcers, Chilblains, and all Eruptions of the Skin. Price
50 cents; by sending 60 cts. to Weeks & Potter, Boston,
Mass., will be forwarded free by mail. For sale by all
Druggists. 497-522o

GOLD PENS!

The Johnson Pen is acknowledged by all who have used
them to be the best pen for the least money of any in use.
They are made of 14 karats Fine Gold and warranted for
one year (written guarantee when required). Pens in Solid
Silver Extension Cases, $1 75; Ebony Slide Holders,
$1 75; Rubber Reverse Holders, $2 50; Telescopic Extension Cases, $3 50; Duplex Silver Cases, $4; Ebony
Holders and Morocco Boxes, $1 50; Pens repointed, 50
cents each. Pens sent by mail, postage paid. Send for
Circular.

I. S. JOHNSON, Manufactory and Office,
496-50o   15 Maiden Lane, N. Y.

Starr's Repeating Four-Shooter.
SAFEST POCKET-PISTOL MADE.

The Advantages of this arm are:
1. It is the only pistol that can be placed in position to receive the Cartridge with one hand.
2. It can be loaded in the dark quicker than any other Pistol, and with no danger.
3. It carries a heavier Cartridge than any pistol of the same size.
4. It is the safest Pistol to load, shoot or carry, as a premature discharge is impossible.
5. It has less parts than any other Pistol, and is less liable to get out of order.
6. It shoots the ordinary metal Cartridge that can be purchased anywhere.

TO BE FOUND AT THE PRINCIPAL GUN AND HARDWARE DEALERS.

MERRILL PATENT FIREARMS CO.,
o   Baltimore, Md.

ARCANA WATCH.
An Elegant Novelty in Watches.

The cases of this watch are an entirely new invention,
composed of six different metals combined, rolled together and finished, producing an exact imitation of
18 carat gold, called Arcana, which will always keep its
color. They are as beautiful and durable as solid gold,
and are afforded at one-eighth the cost. The case is
beautifully designed, with Panel and Shield for name,
with Finest Push Pin, and engraved in the exact style
of the celebrated Gold Hunting Levers, and are really
handsome and desirable, and so exact an imitation of
gold, as to defy detection. The movement is manufactured by the well-known St. Jimer Watch Company of
Europe, and are superbly finished, having engraved pal-
lets, fancy carved bridges, adjusting regulator, with
gold balance, and the improved ruby jewelled action,
with fine dial and skeleton hands, and is warranted a
good timekeeper. These watches are of three different
sizes, the smallest being for ladies, and are all Hunting
Cases. A case of six will be sent by mail or Express for
$25. A single one sent in an elegant Morocco Case for
$5, will readily sell for three times their cost. We are
a sole agents for this watch in the United States, and
one are genuine which do not bear our Trade Mark.
Address   DEVAUGH & CO., Importers,
                        15 Maiden Lane, New York.

BASHFULNESS.—How to overcome it. See
PHRENOLOGICAL JOURNAL, January No., 30 cents.
500-3o   FOWLER & WELLS, 389 Broadway, N. Y.

USE THE BEST!
IT WILL CERTAINLY
DESTROY MOTHS. Now

A BEAUTIFUL ENC
Case, Lever Cap, S
Hands, "English A
with an accurate
Day of the Week, M
sent free, by mai
case, WITH A BEAU
A neat SILVER
Miniature Calenda
Sent free by ma
only $7.
English and Am
Watches of all des
Address CHAS.
o

Violette

This new, fragra
last and best) is su
It has no equal.
W. H.
55 Rue d'Enghie

Cured by Bates's A
&c., address H. C.
eowo

WILLIA
M.
GOI
Grand
P

nt on application
A full assortme
ments at

For Hardenin

Cleansing, Beau
Purifying and S
venient, efficacio
the world has
Sold by
where—

208 1
Agents w
o

Address FRANK

Here at last is a definitive book on the first successful type of multishot firearms. Lew Winant, a well known collector and student of arms, brings to the collecting fraternity a treasury of information about all true and quasi types of pepperboxes within the covers of a single volume.

In this book the arms collector will find descriptions of a comprehensive collection of pepperboxes. Students may trace the development of the weapon from its first cumbersome crude form in the 17th century to the sleek, easily concealed, deadly killer produced just before the turn of the present century.

The variations in form and function of this dramatic arm are carefully noted through its entire development, and particular emphasis is given to its popular use in the 1800's. In this period pepperboxes were the arms that sailed with salty captains around the Horn, the "hide-outs" of the foggy water-

PEPPERBOX FIREARMS

fronts, the "lead sprayers" of the Vigilantes, the "palmers" of the gambling hells, the "big-uns" of the prospectors. These firearms, which were yesterday's protectors of life and property, are today enthralling specimens for the collector's attention.

PEPPERBOX FIREARMS contributes much to the story of man's eternal striving for "fire-power"—for those extra shots to "outgun" adversaries. A galaxy of these multishots are pictured for the approval, delight and interest of all. Every sort is shown—American and foreign, those of various ignition forms, many shapes and sizes, including those fired from the shoulder and those that are "cased" with full complement of accessories.

Twelve authoritative and enjoyable chapters adequately cover the history of this fabulous weapon. There are ninety-two plates showing hundreds of types of pepperboxes, precursors in many ways of our modern automatic weapons. The collector and scholar will find in PEPPERBOX FIREARMS a wealth of facts aligned in precise and orderly sequence covering an important facet of armslore.

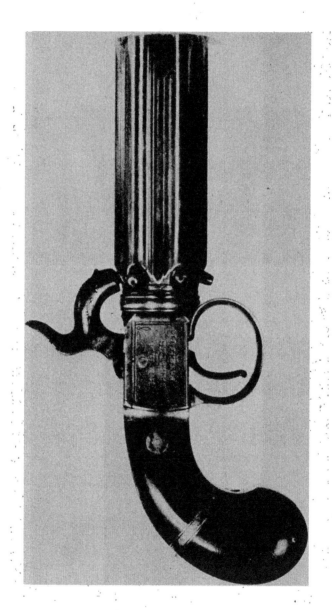

Frontispiece B. & B. M. Darling pepperbox. The first patented American pepperbox.

PEPPERBOX FIREARMS

by Lewis Winant

New York
GREENBERG : PUBLISHER

FOREWORD

Many people helped me with this book, not only with material but with the benefit of their knowledge on debatable points. Other dealers told me to look though their stocks and to help myself. Collectors whom I asked to send me prized pieces to study were very generous in letting me have them.

I regret that it is impossible to supply a detailed listing of the assistance given me. Doctor Prentiss D. Cheney gave me full co-operation, supplying photographs and detailed descriptions and helping to resolve problems; so did Mr. Frank Horner by sending many fine pieces and offering valuable suggestions. Mr. G. Charter Harrison, Jr. sent photographs and told me to use anything I wished that had appeared in *The Gun Collector*. Mr. Paul Westergard found important material for me. These are only a few instances.

I hesitate to list those who assisted in one way or another, because I may inadvertently omit someone who gave important help. Those whose names come to mind are:

Robert Abels, Prentiss D. Cheney, P. J. Federico, Anthony A. Fidd, W. R. Funderburg, G. Charter Harrison, Jr., Leonard Heinrich, John Hintlian, Frank R. Horner, George N. Hyatt, W. G. C. Kimball, Joseph Kindig, Jr., William M. Locke, R. H. McLeod, Arnott Millett, A. D. Mowery, W. Keith Neal, John E. Parsons, Gordon Persons, Glode M. Requa, Ray Riling, Robbins H. Ritter, William Sartain, James E. Serven, Sam E. Smith, Joseph L. Spicer, Paul J. Westergard, Harold G. Young.

The name of the owner of an illustrated piece appears below the illustration unless the piece is or was owned by me. Pepperboxes shown on the jacket are from my current stock.

The photographs were taken by several men, at various times and places. The size of the picture often is not related to the size of the actual gun, but usually the size of the gun is stated in the caption.

Contents

	FOREWORD	
I.	WHAT IS A PEPPERBOX?	7
II.	VERY EARLY PEPPERBOXES	11
III.	THE DARLING PEPPERBOXES	20
IV.	THE ALLEN PEPPERBOXES	27
V.	OTHER AMERICAN PERCUSSION PEPPERBOXES	46
VI.	THE SHARPS PEPPERBOXES	78
VII.	OTHER AMERICAN CARTRIDGE PEPPERBOXES	87
VIII.	EUROPEAN PERCUSSION PEPPERBOXES	99
IX.	EUROPEAN CARTRIDGE PEPPERBOXES	134
X.	? ? PEPPERBOXES ? ?	149
XI.	PEPPERBOX SHOULDER PIECES	172
XII.	CASED SETS AND ACCESSORIES	177
	BIBLIOGRAPHY	187

The front endpaper has an advertisement of a Starr which appeared in *Frank Leslie's Illustrated Newspaper*, May 6, 1865, and an advertisement of a Marston which appeared in the *Army and Navy Journal*, July 14, 1866. The back endpaper has an advertisement of Elliot's Pocket Revolver which appeared in *Harper's Weekly*, April 12, 1862, and an advertisement of Elliot's New Repeaters which appeared in *Frank Leslie's Illustrated Newspaper*, September 12, 1863.

Chapter 1
WHAT IS A PEPPERBOX?

THE PEPPERBOX was the first type of multishot firearm to achieve widespread popularity and sale.

For the present purpose a pepperbox is defined as *any hand firearm with three or more barrels encircling a central axis, firing shots successively with only one striker.*

Firearms conforming to this definition were made over 350 years ago. Pepperboxes such as the Holy Water Sprinkler were made before 1600. The Shatuck Unique is a twentieth-century product.

Hardly any student will wish to restrict this definition further by including only pepperboxes with revolving barrels. To do that would eliminate guns made by Robbins & Lawrence, which are always classed as pepperboxes.

Some very interesting multiple-barrel hand arms which are not generally called pepperboxes, but which satisfy all conditions of a previous and more elastic definition, will be described in Chapter 10.

A pepperbox differs from a revolver in that it has no one directing barrel for its shots. In a modern revolver, all shots pass through a single barrel; a pepperbox has a separate barrel for each shot.

The conventional double-action, percussion-cap pepperbox lock has a trigger which raises the hammer against the com-

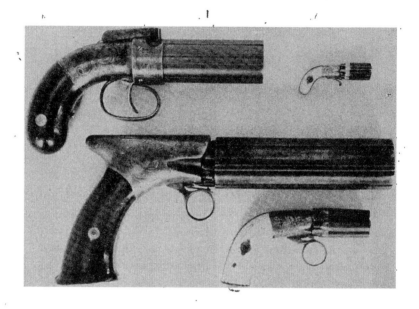

Fig. 1 Group showing relative sizes of pepperboxes.
(THE MINIATURE IS FROM THE DR. FUNDERBURG COLLECTION.)

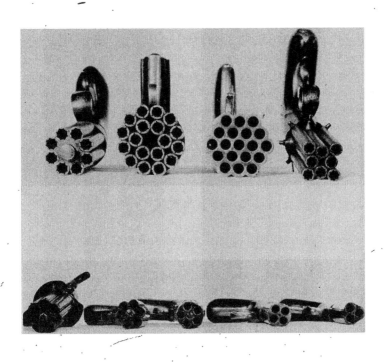

Fig. 2 Fourteen pepperboxes showing muzzles.
(SIX ARE FROM THE FRANK HORNER COLLECTION.)

pression of the mainspring until the sear disengages and the hammer is free to drive down on the cap. Usually the trigger action also revolves a cylinder by means of a pawl turning a ratchet fixed to the cylinder. In most cases, in pepperboxes with revolving cylinders, a small bolt, also operated by the trigger, locks the cylinder at the moment of firing. Nearly always a trigger spring is provided to return the trigger automatically to its original position after each discharge.

Single-action pepperboxes with mechanically turned cylinders usually have the hammer linked to a pawl which engages a ratchet on the cylinder. Some double-action and some single-action pepperboxes have manually revolved cylinders. Such pepperboxes, and others with exceptional mechanisms, will be described in detail.

All that is necessary to fire a double-action pepperbox is to pull the trigger. To fire a single-action pepperbox, the hammer must be cocked separately before the trigger is pulled.

Figure 1 shows the relative sizes of miniature, small, medium, and large percussion-cap pepperboxes. Figure 2 shows muzzle views of some of the many different barrel arrangements found on percussion-cap pepperboxes with from three to eighteen barrels, and on cartridge pepperboxes with from four to twelve barrels. The individual pieces will be described in later chapters.

Percussion-cap pepperboxes are considered antique in the usually accepted sense. Nearly all are ninety years old, and most are fully one hundred. Of course, under United States law no gun made before 1830 is dutiable. Though a very few percussion-cap pepperboxes were made in Europe before 1830, it might be very uphill work to convince the Customs that any percussion pepperbox is not dutiable.

Some thousands of percussion-cap pepperboxes still survive. A small collection of the more common models is easy to obtain, but a collection of the rare makes and models is difficult to acquire.

Chapter 2
VERY EARLY PEPPERBOXES

WE CAN ASSUME with utmost assurance that the first gun was a single-shot, but we know there were very early guns designed to shoot more than once without reloading.

The problem of repeaters was approached in many ways. Guns were made with several barrels, sometimes fixed and sometimes turning; some had cylinders in the style of modern revolvers; many were made to take superposed loads in a single barrel; a few had complicated magazines with intricate devices for mechanical cocking and priming. The pepperbox is just one form of multishot weapon.

The Holy Water Sprinkler, pictured in Figure 3, is an elaborate form of the first pepperbox. In most cases the Holy Water Sprinkler, or Morning Star, is simply a medieval mace fitted with spikes. This very rare specimen has four barrels, or hand cannon, with individual touch holes. There is a cover for the muzzle and there are separate covers for the touch holes. When these covers are closed, it is not apparent the weapon is a firearm. This ingenious German piece is of wood with fancifully engraved bone inlays. The four barrels, each 9 inches long, are separated by and covered with wood to form a cylinder which is held with four iron bands, two of which have six spikes each. The piece is shown with the muzzle covered and with one touch hole cover drawn back, exposing a vent. The muzzle cover, with its 4-inch spike, is hinged and may be opened by pressure on a spring. The striker for this first pepperbox was a "match" or "coal" which was touched to powder in a touch hole. The covers for the touch holes are of engraved bone and slide in grooves. The ramrod is contained in a trap in the staff, whose end has a small sliding cover. The first repeating hand firearm was quite probably made by binding together several "cannones," or hand cannon, as was done in this Holy Water Sprinkler. From that premise follows the inference that the pepperbox was the first multishot hand firearm.

The firing of wheel lock and flint guns requires that a spark be generated by friction in the open, in priming powder outside the chamber. A safe, dependable repeater with this ignition was impossible. The possibility of sparks getting to more than one chamber always existed and increased as the fittings were worn with use. A multiple explosion could be disastrous both to the gun and the shooter.

Nevertheless, at least one pepperbox with wheel lock ignition was made. There was such a piece in the Historical Museum in Dresden. And there was a demand for flint repeaters, despite their faults, but such guns were expensive and the market was limited to the wealthy.

One of the earliest of flint pepperboxes is that shown in Figure 4. This was made in Utrecht, Holland, by Jan Flock, who was making guns well before 1700. Note the very early hammer with cross-slotted screw, and the shape of the grips, which are of wood with an ivory cap. After pressing back the trigger guard, the four barrels are turned manually until a catch locks the cylinder with a vent opposite the pan.

The flint pepperbox by Jover, London, Figure 5, is a finely engraved piece made for presentation in the Orient. The furniture and the ornamental inlays are of silver, except for a halfmoon, surrounded by stars, all of gold. The three barrels, gold inlaid with flowers and leaf designs, have front sights only. (The use of rear sights on early guns was usually reserved for shoulder weapons.) The barrels are turned by hand without a separate release mechanism. The pan, both top and bottom, is of thick gold and the touch holes are set in thick gold that completely encircles the cylinder breech.

A particularly fine flint pepperbox is that pictured in Figure 6. This is one of a pair made by Mathias Bramhofer of Augsburg. The two flint pepperboxes just described are each fitted

Fig. 3 Holy Water Sprinkler—36″ overall—four-shot—.50 caliber. (JOSEPH KINDIG, JR., COLLECTION.)

with only one frizzen and pan, necessitating priming after each shot. This one has a separate frizzen and pan for each of the four barrels, so that priming between shots is obviated. To fire successive shots it is necessary only to cock the hammer, give a quarter-turn to the cylinder, and pull the trigger. The mechanism for turning the cylinder is unique. The trigger guard is not movable, its front end being pivoted to the frame, but it has a floating rear end which moves into the frame, so that pressure on the guard withdraws the bolt that locks the cylinder. Shown fitted in the muzzle is a combination accessory. The visible part is a small screwdriver; the hidden part consists of four hollow tampions fastened to a round plate. These serve as powder measures. When filled with powder, these containers are inserted as a unit in the muzzle, with the gun pointing down. When the gun is inverted, all barrels are loaded with powder.

A quite different type is shown in Figure 7. This type was chunky, clumsy, and heavy, but it was well made and reasonably dependable. J. N. George in *English Pistols and Revolvers* suggests that two to three hundred were made in the period 1780-1800. He adds that Henry Nock was probably the manufacturer of nearly all of this type that were made, though they bear the names of many gunsmiths. This piece has seven barrels, but fires only six times. Six barrels are screwed into a solid breeching and encircle the seventh barrel, screwed into the center. The barrels are screwed in or out with a muzzle key. Cut into the solid breeching are six pans, with vents to the barrels. The powder chambers of barrels numbered six and seven are connected, so that when barrel six is on top and fired, barrel seven goes off simultaneously. These guns usually have a slide which permits locking the cylinder. On this particular piece the slide is arranged to lock the cylinder only when barrel six is in firing position. This arrangement gives support to the belief that the twofold shot was intended to be the first one. A contrary belief is that the double blow was to be reserved for a "finisher." The breeching is attached to the frame by a spindle and is fastened by a thumb screw. It is held with a pan in firing position by a spring catch working on the spindle. The barrels are turned

Fig. 4 Jan Flock, Utrecht—10½" overall—four-shot.
(JOSEPH KINDIG, JR., COLLECTION.)

Fig. 5 Jover, London—15" overall—three-shot—.50 caliber.
(JOSEPH KINDIG, JR., COLLECTION.)

manually against this spring pressure which may be regulated by a second thumb screw. The pan of the barrel in firing position has a cover that is part of the frizzen. The other pans are covered by a sleeve over the breeching. This sleeve is tightly fitted to guard against fire passing to powder in an adjacent pan. This piece has more appendages than we expect in a firearm: two thumb screws, a sliding cylinder bolt, and also a sliding safety. This safety, which is also found on the three-and four-barrel pieces to be described next, locks the frizzen closed and also the hammer at half-cock.

Figure 8 is of a pepperbox commonly called a "three-barrel tap action flint pistol" that is not often classed as a pepperbox. Actually, as it has three barrels encircling a central axis, firing shots successively with a single striker, it is, logically, a species of the genus pepperbox. There is a lever on the left of the frame which revolves a cut-off on the water tap or faucet principle. When this lever is fully raised, the left barrel on top may be discharged. Pressing the lever partly down exposes the vent to the right top barrel so that it may be fired. Pressing the lever fully down opens the vent to the bottom barrel. When the trigger guard is pulled back, the bayonet snaps out violently and is held in position by a spring-controlled catch.

A four-barrel pepperbox that combines the turning tap action with a sliding pan cover is shown in Figure 9. The manipulation of the levers, with the order of fire, is:

(1) Tap action lever on left, raised. Sliding pan cover lever on right, forward. Left top barrel fires.
(2) Turn down lever on left. Left lower barrel fires.
(3) Pull back sliding lever on right. Right top barrel fires.
(4) Raise lever on left. Right lower barrel fires.

Priming between shots was not necessary with these tap action pepperboxes. The revolving cut-offs were hollowed to contain priming for all shots.

Between shots on this and on all the other flint pieces described, it is of course necessary to draw the hammer to full cock and to close the frizzen. The bayonet on this four-barrel piece is released by pulling a small trigger under the frame.

Fig. 6 Mathias Bramhofer, Augsburg—12" overall—four-shot—
.39 caliber. (METROPOLITAN MUSEUM OF ART COLLECTION.)

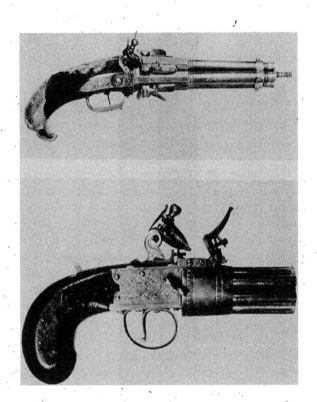

Fig. 7 Wood, York—8¾" overall—seven barrel—.33 caliber.

Some of the most skilled and most patient artisans the world has known were early gunmakers, but until the invention of the percussion cap they could not produce a wholly reliable repeating gun. When Alexander Forsyth applied a detonating lock to a gun, the great obstacle was soon to be overcome. In a few years came the percussion-cap development. The road was open for repeaters and many gunsmiths tried their hand at pepperboxes.

The primitive percussion-cap pepperboxes are here considered as being crude pieces, made experimentally in small numbers. They employed no new principles. Either a number of barrels were brazed together, or a solid cylinder was drilled with several holes to serve as barrels. The barrels were vented and nipples screwed in. All early pepperboxes classed as primitives had barrel groups revolved manually. The turning might or might not require separate release of a catch which when engaged locked the barrels with a nipple under the hammer. The usual method was the same as that frequently employed on over-and-under Kentucky rifles—pulling back the trigger guard to release the catch.

The operation of a primitive percussion pepperbox requires three separate manipulations: (1) turning the cylinder, (2) cocking the hammer, (3) pulling the trigger. The operation of the first model Darling pepperbox, described in Chapter 3, requires two manipulations: cocking the hammer and pulling the trigger. The operation of the Allen, described in Chapter 4, requires only the pulling of the trigger.

A somewhat unusual primitive percussion cap pepperbox is shown in Figure 10. It is unmarked but is believed to have been made in Pennsylvania. The construction is of iron, except for a top plate and trigger guard of brass and a one-piece wood grip. The six barrels are welded to a core that is drilled for a spindle extending from the frame. A nut on the end of the spindle secures the barrel group. The maker found a novel and effectual method of locking the cylinder. He used a second trigger which moved a bolt in and out of a cylinder recess. This piece is primitive in every sense.

Other early pepperboxes with primitive features—those of

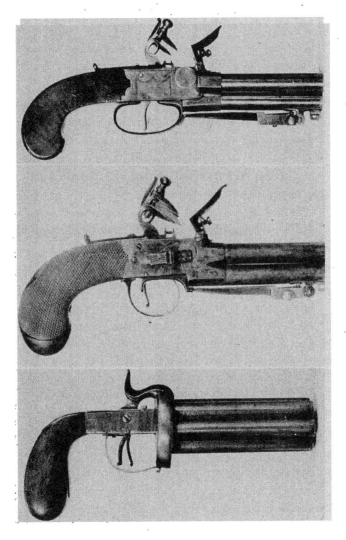

Fig. 8 (*top*) Wallis, Hull—8¾" overall—three-shot—.38 caliber.
Fig. 9 (*center*) Twigg, London—8¾" overall—four-shot—.38 caliber.
Fig. 10 (*bottom*) Unmarked—8¼" overall—six-shot—.40 caliber.

regular production and those that were finely made and finished —will be dealt with in later chapters.

Chapter 3
THE DARLING PEPPERBOXES

THE BROTHERS Benjamin and Barton Darling were granted a patent for the first American pepperbox.

The history of the operations of the Darling brothers is obscure, but it is known that not many of their patented pepperboxes were made. Perhaps a half-dozen still exist. The Darlings were established gunsmiths in Bellingham, Massachusetts, when their invention was patented. Their first pepperbox was undoubtedly made there. A few more—perhaps twenty-five—were reported made in Shrewsbury, Massachusetts. Then the Darlings settled in Woonsocket, Rhode Island, and for a short time continued to manufacture the pepperboxes they had patented.

Mr. Benjamin Darling has been repeatedly quoted as having maintained until his death at an advanced age that he invented the first American revolver. (The word "pepperbox" was not in vogue at the time of his invention, though it had been used earlier and was to be revived a few years later in America. In the 1830's a pepperbox was a "revolver" or a "rotary pistol.") The Darling patent, dated April 13, 1836, claimed the rotation of the cylinder by cocking the hammer; the Colt patent, which also claimed this mechanical rotation, was dated February 25, 1836—some seven weeks earlier. Conflict with the Colt patent may have been the determining factor in causing the Darlings to cease manufacturing their Patent Rotary Pistol.

The great fire at the Patent Office in December, 1836, de-

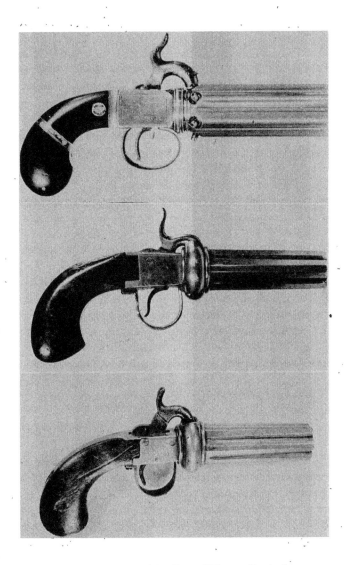

Fig. 11 (*top*) B. & B. M. Darling—6¾" overall—six-shot—.30 caliber.
Fig. 12 (*center*) J. Engh—8¼" overall—four-shot—.28 caliber.
Fig. 13 (*bottom*) Unmarked—8½" overall—six-shot—.31 caliber.
(DR. PRENTISS D. CHENEY COLLECTION.)

stroyed nearly all records. There is almost no information existing either about the contents of early patents or about the routine of filing and recording applications and caveats. Obvious discrepancies are not susceptible to explanation. As an instance, it has not been possible to ascertain why Re-issue #124 obtained by Samuel Colt on October 24, 1848, reads: "Specification forming part of Letters Patent No. 138, dated February 25, 1836 .." The original Colt patent is frequently referred to by collectors as Patent #138, but an inquiry at the Patent Office for a copy of Colt's Patent #138 brought the response that no Colt patent bears the number 138, and that Patent #138 was granted to B. S. Gillespie and covered A Machine for Breaking Ice. The Patent Office states that the 138 considered by collectors as the number of the original Colt patent "... may refer to the annual number which might have been used at that time." For a few years patents were numbered throughout the year, with the No. 1 repeated at the beginning of each year. The present system of numbering dates from July 4, 1836. Whether the annual numbers which antedated the present series were based on the British system of patent application numbers and were given at the time the applications were filed or whether the numbers were assigned when the patents were granted is not known.

It is interesting to note that the first-model Darling pepperbox, shown in Figure 11, is marked B. & B. M. DARLING PATENT 113. The number 113 does not appear to be a serial number. Moreover, collectors are agreed that the total number of these guns made was well under one hundred. Whether this 113 was an annual number in the series of 1836 patents issued prior to July 4, 1836, is not disclosed by records of either the Patent Office or the Hall of Archives.

In any event, neither Colt nor Darling made the first gun with a mechanically turned cylinder. Elisha Collier patented such a gun in England in November, 1818, and was chagrined to be shown a much earlier Swedish gun with a similar mechanism when he submitted his invention to the British Government in 1819. Now it is revealed that the Collier gun was patented by

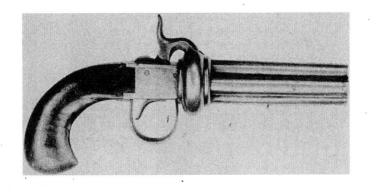

Fig. 14 A C S—8⅞″ overall—four-shot. (WILLIAM M. LOCKE COLLECTION.)

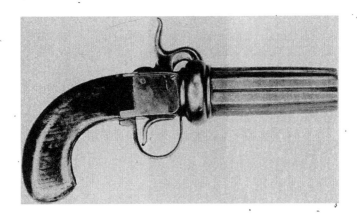

Fig. 15 I E H—8¾″ overall—six-shot. (WILLIAM M. LOCKE COLLECTION.)

Artemas Wheeler in the United States in June, 1818, five months before it was patented in England.*

The Darlings invented their pepperbox when the time was ripe, but theirs is the rarest of American pepperboxes. The gun was costly to produce and even if there had been no question of patent interference, it had no chance to survive in competition with the more rapid fire Allen double-action which quickly followed it.

The six-shot first-model Darling is single-action and made of steel. Cocking the hammer revolves the cylinder by a lever-and-ratchet device. The Darling brothers were not sparing in decoration and refinements. The workmanship is fine; the grips are ornamented, the barrel ribs are channeled, and the frame is engraved.

Advertisements, such as "Darling Patent Rotary Pistol, made and sold wholesale and retail by B. & B. M. Darling, Woonsocket, R. I.," were soon discontinued, and manufacture of the patented pepperbox was abandoned. However, the Darlings did not renounce the gun business. Certain pepperboxes, usually with brass frames and brass barrels, are universally accepted by collectors as Darlings even though they do not bear the Darling name. Most of them have hand-revolved cylinders, but some have been found with mechanically revolved cylinders. All were probably fabricated in Woonsocket. Examples are seldom found; even so, they are not nearly so rare as the patented type, Figure 11.

The brass hand-revolved Darlings are found with no markings as well as with such varied marks as J. ENGH, I. E. H., A C S, A I S. Mr. Sam Smith, in his comprehensive article in the January, 1942, issue of *The Gun Report*, subscribes to the theory that these markings are initials of journeymen who produced the arms under the Darlings' supervision, a theory which is generally accepted. The question of whether "J. Engh" was the mark of Jack Englehart, a noted arms maker of Nazareth, Pennsylvania, in the 1830's, is presented by Mr. Smith merely as a possibility. I have never found any proof that this maker

The Gun Collector No. 35, February, 1951.

Fig. 16 J. Engh—6⅞″ overall—six-shot pin-fire.
(WILLIAM M. LOCKE COLLECTION.)

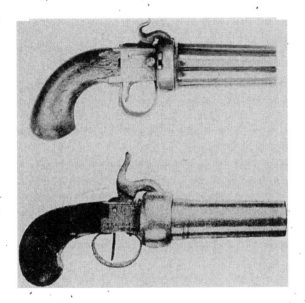

Fig. 17 Schmid Bundelrevolver—Reproduction of illustration in Gerhard Bock's *Moderne Faustfeuerwaffen*.

of Kentucky rifles had any part in making these pistols. There is also no proof that "J. Engh" had been a gunsmith in Europe before working with the Darlings, but there is evidence to support the surmise and to indicate the Darlings were guided by his suggestions. One basis for the supposition is the similarity in design of pepperboxes made in Germany by Schmid and those marked J. ENGH. Another is the fact that at least one pin-fire pepperbox marked J. ENGH exists. The use of pin-fire ignition is completely foreign to American pepperboxes. Of course "J. Engh" may have made this one exotic pepperbox either with or without the Darlings' approval.

It is possible, but not definitely known, that three- and five-barreled brass pepperboxes were made. The four-barrel and to a lesser extent the six-barrel had some success. Not having a complicated mechanism and being made of brass, they could be produced and sold cheaply.

The cylinder on a hand-revolved pepperbox may as a result of wear turn in either direction. When the Darling pepperboxes were made, their cylinders turned clockwise only—though I do know of two, both marked I. E. H., which turn counter-clockwise only. The only other American percussion pepperboxes with clockwise turning cylinders that I can think of are the Bacon and one model of the Pecare & Smith.

A hand-revolved four-shot marked J. ENGH is shown in Fig. 12. Brass is used for barrels, frame, and trigger guard. There is no ornamentation. The cylinder rotates clockwise and the rotation is effected by turning pressure only. Each one-fourth turn permits a plunger to lock the cylinder lightly with a nipple under the hammer.

An example of a six-shot is pictured in Figure 13. Except for very minor differences that are inevitable in guns made by hand, this differs from the four-shot only in size, in having six barrels, and in being quite unmarked.

Figure 14 is of a four-shot marked A C S. This is like the "J. Engh" four-shot, except that all parts are of iron except the nipple shield, which is of brass. The hand-revolved Darlings with iron frames are much scarcer than those with brass frames.

Figure 15 shows one of the rare six-shot brass pepperboxes with a mechanically revolved cylinder. Cocking the hammer rotates the cylinder counter-clockwise. This piece is marked I. E. H.

The pin-fire pepperbox previously mentioned is pictured in Figure 16. This hand-revolved brass pepperbox fires pin-fire cartridges which are loaded through a gate on the right of the frame. This is marked J. ENGH.

Figure 17 is a reproduction of an illustration in Gerhard Bock's *Moderne Faustfeuerwaffen* of a Schmid Bundelrevolver with percussion-cap ignition. In some respects there is marked resemblance in this piece to the brass Darlings. Did Schmid get ideas from a Darling pepperbox, or did an immigrant gunsmith influence the Darlings in designing a low-cost pepperbox? We just don't have the answer.

Chapter 4
THE ALLEN PEPPERBOXES

THERE ARE two groups of Allen pepperboxes: those bearing the names of Allen companies, and those made by the Allen companies for dealers and bearing such names as Bolen, J. Eaton, Lane & Read, Spies, Tryon, Warren & Steele, Young & Smith. Pepperboxes of this second group are stamped "Allen's Patent."

The first category includes those pepperboxes manufactured at Grafton, Mass. (1837-1842), and at Norwich, Conn. (1842-1847), all stamped ALLEN & THURBER. It also includes those manufactured at Worcester, Mass. (1847-1865), variously stamped ALLEN & THURBER, ALLEN THURBER & Co. and ALLEN & WHEELOCK. After 1865 the firm continued under the name of Ethan Allen & Co., but pepperboxes were not made after the Allen & Wheelock era.

Ethan Allen was a gunsmith in Bellingham when the Darling brothers lived there, but just befqre the Darling patent was secured, Allen moved to Grafton. There he set up a shop with his brother-in-law, Charles Thurber, and there they made pistols and later pepperboxes. The Allen companies were always a family affair. T. P. Wheelock, of Allen & Wheelock, was another brother-in-law. Messrs. Forehand and Wadsworth were sons-in-law and partners of Ethan Allen after Wheelock's death.

Ethan Allen was a pioneer in the transition from handmade to machine-made and interchangeable parts. He probably produced more different kinds of guns—everything from cane guns to fowling pieces and Fourth of July pistols—than any other manufacturer, but we are concerned here only with pepperboxes.

The Allen pepperbox was the first American double-action pepperbox and it was a big success. Trigger action rotated the cylinder and raised the hammer. As quickly as the trigger could be pulled fully back, the hammer was released and the gun fired.

For a dozen years and more after the Colt revolver was first made, sales of Allens far outstripped those of Colts. In 1847, according to the Connecticut Historical Society report, Captain Walker wrote Colt from Washington that ". . . nine men of Ten in this City do not know what a Colt Pistol is and although I have explained the difference between yours & the six barrel „Pop Gun,, that is in such general use a thousand times they are still ignorant on the subject . . ."

It will not detract from the renown of the manufacturer who made the first immediately successful American multishot firearm to correct two major mistakes about him that persist in print. The first error lies in associating the firearms manufacturer with the Revolutionary War officer; the second, in referring to Ethan Allen's 1837 patent as a patent for a pepperbox. Misunderstanding became so rooted that a current dictionary incorrectly defines a pepperbox as "a popular name for a pistol invented by Ethan Allen about the time of the American Revolution."

Several independent investigations fail to disclose any relationship between Ethan Allen, the firearms manufacturer born in Bellingham, Mass. in 1808, and Ethan Allen, the hero

Fig. 18 Allen Patent Model—6¾" overall—six-shot—.30 caliber.
(SMITHSONIAN INSTITUTION COLLECTION.)

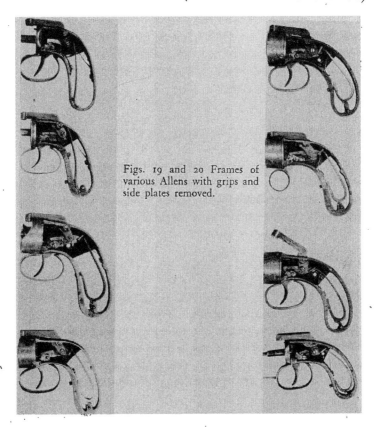

Figs. 19 and 20 Frames of various Allens with grips and side plates removed.

of Ticonderoga, who died in 1789. Family relationship between the two men is easily possible, but the Massachusetts Historical Society in Boston states that to its knowledge the fact of relationship has not been established.

The 1837 Allen patent related only to a method of both raising the hammer and driving it down with one pressure of a trigger. It pictured and described the double-action lock mechanism as being for a single-shot pistol.

The Allen pepperbox could be put in action fast. It met the need for a reliable weapon of defense at close quarters and was welcomed by travelers—and emigrants particularly. In those days men rarely traveled any distance unarmed; they were familiar with and knew how to handle a firearm, and they usually kept one under the pillow. Thieves entered at their peril. The Allen imparted a feeling of security not given by a single-shot pistol, and its rapidity of fire over the single-action revolver outweighed the revolver's greater accuracy in the minds of men looking for a weapon for emergency use. A sudden emergency gave no time for deliberate aim.

The Allens were very popular with the Forty Niners. Allens reached California by the cross-country route, by way of the Isthmus of Panama—over, not through, in those days—and by the long way 'round the Horn.

The pepperbox was the fastest shooting hand gun of its day. Many were bought by soldiers and for use by state militia. Some saw service in the Seminole Wars and the War with Mexico, and more than a few were carried in the Civil War. A report by the American Ordnance Bureau, listing firearms that had been used regularly by the U. S. Army, from "Earliest Times to 1903," mentions "Revolving pistols—pepperbox—percussion." In *The History of the United States Army* by William Addleman Ganoe, mention is made of the use of Allens in battles as late as 1857 between U. S. Cavalry and the Cheyennes. But because of its small bore, short range, and lack of accuracy, the pepperbox was by no means as satisfactory as a revolver for military use. It could not be properly aimed. The heavy trigger pull and the turning of the barrels disturbed the aim. Furthermore, the top

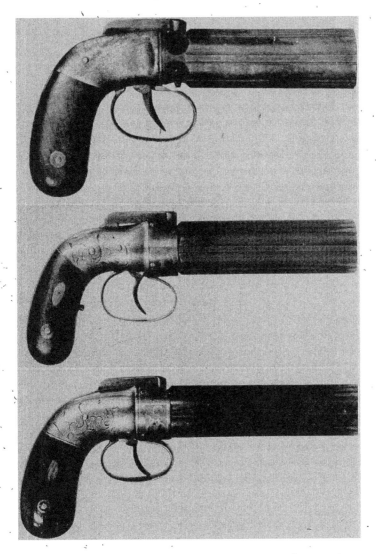

Fig. 21 *(top)* Allen & Thurber, Grafton—6¾" overall—six-shot—.31 caliber—serial #27.

Fig. 22 *(center)* Allen & Thurber, Norwich—8" overall—six-shot—.33 caliber—serial #1.

Fig. 23 *(bottom)* Allen & Thurber, Norwich—9¼" overall—six-shot—.36 caliber—serial #45.

hammer was placed directly in the line of sight.

With either a revolver or a pepperbox, it was a disconcerting but not uncommon experience to have all six barrels go off in unison. The partitions, placed between nipples to prevent fire from one barrel reaching another, were not always effective. The shield around the nipples was designed to prevent dislodging of caps and injury to nipples, as well as to protect against dampness and to minimize misfires. Some pepperbox users thought a nipple shield increased the danger of multiple explosions, because flame might be carried under it. The British and Continental makers were inclined to dispense with these enclosing shields. The absence of shields on some Allens is undoubtedly due to customer preference, but generally the shieldless Allens are those lacking the other refinements of shielded Allens. Since the best of the Worcester-era Allens have shields, it is assumed that Allen never believed the shield was a detriment.

Mark Twain showed a surer knowledge of firearms than students expect in fiction. In *Roughing It* there is the amusing tale of Bemis' shooting a tree-climbing buffalo with an Allen. In telling of his anger when his veracity was questioned, Bemis said, "I should have shot that long gangly lubber they called Hank if I could have done it without crippling six or seven other people—but of course I couldn't, the old Allen's so confounded comprehensive."

Duels were fought with all manner of firearms, but the rigidly supervised "affairs of honour" between men of high position were settled with costly weapons of high quality, designed especially for duelling. Men who settled their disputes with pepperboxes were unlikely to have their names long remembered. The only such duel I find recorded is a meeting between two ladies in Buffalo, New York. They used Allens, but the authorities stopped the duel by arresting the participants.

The Allens were made four-, five-, and six-shot; in barrel lengths from just under 3 inches to almost 6 inches, and in calibers from about .28 to .40, with the largest guns having the biggest bores. These largest ones were favored by the military. They were commonly called Dragoons and were usually fitted

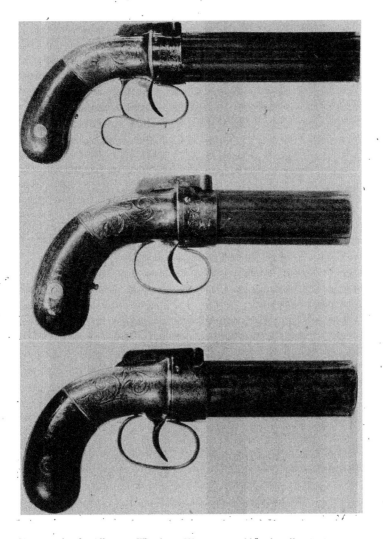

Fig. 24 (*top*) Allen & Thurber, Worcester—9¼" overall—six-shot—.36 caliber—serial #96.   (GLODE M. REQUA COLLECTION.)

Fig. 25 (*center*) Allen & Thurber, Worcester—7¼" overall—six-shot—.31 caliber—serial #159.

Fig. 26 (*bottom*) Allen & Thurber, Worcester—7" overall—six-shot—.31 caliber—serial #82.

with a spur on the trigger guard and sometimes with a belt hook.

Factory records of Allen models do not exist. The number of models of Allen pepperboxes may be considered as from perhaps twenty to upwards of one hundred, depending upon how minute are the modifications the student wishes to count as model changes. Allens examined only externally reveal many small differences: the shapes of the grips, the positions of plate screws and tension screws, the presence or absence of nipple shields and silver ornaments. In addition to the bar hammer type there is the concealed hammer type. And each of these types was made both with conventional triggers and with ring triggers.

The gun numbers are called serial numbers in the captions for the Allens illustrated here, but the numbers are not serials in the sense of indicating sequence of manufacture. The Allen stamped 27 was made before the one next pictured which is stamped 1. The one stamped 37 was made much later than the one stamped 90.

We know an Allen stamped "Worcester" is later than one stamped "Norwich," and that the latter is later than one stamped "Grafton." It is well not to attempt to date or place an Allen more accurately in the sequence, either by reference to one or two features of construction, or by the markings on barrel or hammer except in rare cases. (As far as I know, no Grafton Allen bears the patent date. A Norwich Allen that does not bear the patent date may be assumed to be very early in the Norwich period. Allen moved to Norwich in 1842, the year in which it became mandatory to mark a patented article with the year of the patent.) Allen's second patent, his first for a pepperbox, was #3998, granted in 1845. But Allen pepperboxes made after 1847 and embodying the 1845 features will be found stamped with the 1837 patent marking. The 1845 patent covered an improved method of mechanically turning the cylinder and also a feature of doubtful value that permitted using the weapon either single- or double-action. Figure 18 shows the Smithsonian Institution's patent model. The hammer has spurs to allow easy cocking for single-action use. Cocking this spur hammer did not rotate the cylinder, and this form of hammer

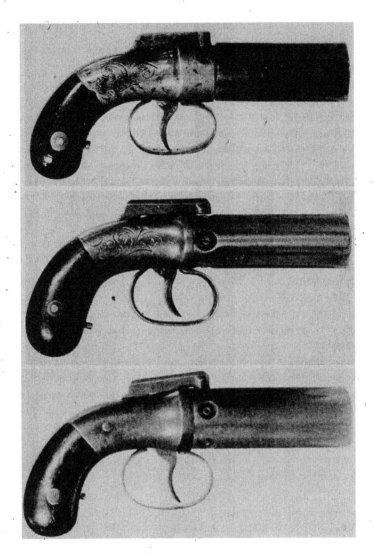

Fig. 27 (*top*) Allen Thurber & Co., Worcester—6¼″ overall—five-shot—.32 caliber—serial #780.

Fig. 28 (*center*) Allen & Wheelock, Worcester—5¾″ overall—four-shot—.31 caliber—serial #480.

Fig. 29 (*bottom*) Allen & Wheelock, Worcester—6″ overall—five-shot—.31 caliber—serial #468.   (HAROLD G. YOUNG COLLECTION.)

was soon abandoned. However, the internal mechanism that permitted this dual use of the hammer continued off and on through the Worcester period, on guns with the 1845 markings as well as those with 1837 markings.

Figures 19 and 20 show the frames of several Allen pepperboxes with grips and side plates removed. The pepperbox shown with hammer raised is a Worcester Allen, but it bears the 1837 patent date. The hammer was pulled to full cock, where it rests, without the trigger being drawn back or the cylinder rotated.

A general trend of economy in the sequence is evident. Refinements appear and disappear. As time went on, the silver ornaments in the grips were discontinued, the barrels were made without ribs, the nipple shields were eliminated, engraving was discontinued.

The omission of "Allen's Patent" from any pepperbox made by Allen was probably an oversight. This stamping is usually on the hammer, but on the Grafton model illustrated it is on the frame. The models with concealed hammers had this marking on the barrel.

Various makers supplied minor parts, such as trigger guards, made to Allen's orders. This may account for the finding of parts identical with Allen parts on Allen copies, the imitators having bought them from the same source. (The imitators were rivals who copied the appearance but did not infringe the patents. They will be discussed in the next chapter). Cases and accessories such as molds and flasks were probably all bought from other makers. Illustrations of various cased sets and accessories will be found in Chapter 12.

Allens are most often found with wood grips of smooth walnut, but they were supplied at extra cost with grips of ivory, horn, or silver. Usually the frame and shield of an Allen is engraved with a quickly executed design. As this was all hand work, no two designs were exactly alike. Occasionally on special order the engraving was elaborate and comparable to the fine work found on European pieces.

Some Allens had rear sights cut in the hammer near the fulcrum, and a few exist with front sights.

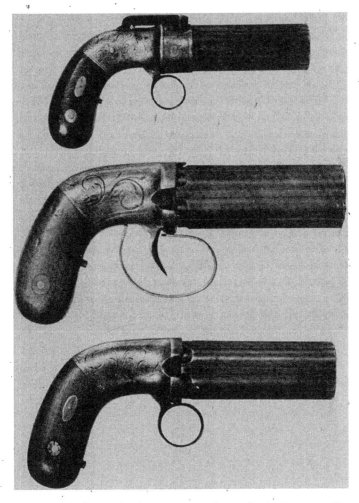

Fig. 30 (*top*) Allen & Thurber, Norwich—7″ overall—six-shot—.31 caliber—serial #90.

Fig. 31 (*center*) Allen & Thurber, Worcester —6½″ overall—six-shot—.31 caliber—serial #128.

Fig. 32 (*bottom*) Allen & Thurber, Worcester—6½″ overall—six-shot—.31 caliber—serial #37.

I am giving considerable attention to variations of construction in the following descriptions of illustrated Allens, because collectors who specialize in pepperboxes are apt to give more heed to minute differences in the percussion Allens—and the cartridge Sharps—than to the mutations in other makes. There are many Allen and Sharps pepperboxes, but of most other makes there are few. Of course, some of the Allen models are in very limited supply.

Figure 21 is of a Grafton Allen. Studious examination of it reveals many differences between it and other top-hammer Allens. It has the flat mainspring and nearly right-angle grips that prevailed on the early Allens. This piece is assumed to be one of Allen's earliest productions by reason of its having other features that were dropped during the Grafton era and never brought back on the Norwich and Worcester models. The grips on this pepperbox are in four pieces and are not held with grip pins. The mechanism was made without the pitman designed to lock the cylinder at the moment of firing. After the pitman was adopted at an early date, it was never discontinued.

This Grafton Allen has removable nipples screwed into the powder chambers. It is somewhat remarkable that this feature was not retained. In all other Allens illustrated here, the nipples are integral with the barrels and made with a milling machine.

The barrel ribs of this piece are channeled. Some Norwich Allens have channeled ribs, but on later Allens as long as ribs were used they were flat. The frame is not engraved and was made without nipple shield. Engraving and nipple shields appeared at an early time in the manufacture of Allens, to be abandoned only at the end. Silver ovals in the grips were about the first refinements to appear and also the first to vanish. This Grafton Allen seems to have preceded their introduction.

Most Allen top-hammer pepperboxes have the stamping "Allen's Patent" on the hammer, sometimes on top and sometimes on the left side. This one has ALLEN's PATENT on the frame. It has ALLEN & THURBER GRAFTON MASS. stamped on the hammer. The firm name is usually on the barrel, and the barrel marking may be in one, two, or three lines. Each Allen

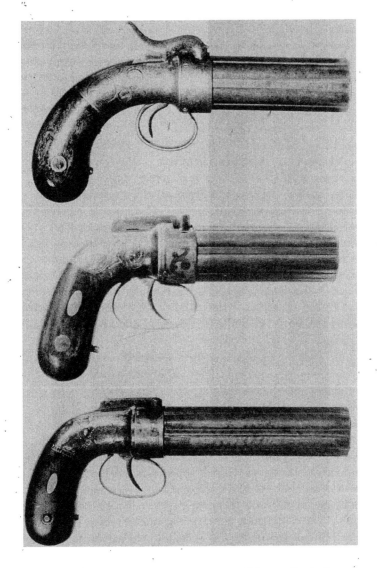

Fig. 33 (*top*) Allen & Thurber, Worcester—7½″ overall—six-shot—.32 caliber—serial #144.

Fig. 34 (*center*) J. G. Bolen—6¾″ overall—six-shot—.31 caliber.
(PAUL J. WESTERGARD COLLECTION.)

Fig. 35 (*bottom*) A. W. Spies—9¼″ overall—six-shot—.36 caliber—serial #12.
(ARNOTT MILLETT COLLECTION.)

pepperbox has several parts numbered alike. Usually the numbers cannot be seen without removal of the grips. On this Grafton Allen the number 27 is stamped on the outside of the frame as well as inside. This may have been, but probably is not, the twenty-seventh Allen pepperbox manufactured. The number is not supporting evidence.

The Allen pictured in Figure 22 has the numeral one on various parts, but it was not made until after Allen & Thurber left Grafton. This has all the refinements, such as a nipple shield fastened with screws and ornamented with stamped scroll work, silver ovals in the grips, and engraved frame. It has the improved internal construction with the loop mainspring. The trigger spring is now omitted and the spring operating on the revolving mechanism is now attached to the lower part of the strap, which is much narrower than the strap on the Grafton product. The location of stampings does not conform to that on most Allens. ALLEN'S PATENT 1837 CAST STEEL is on the barrel and ALLEN & THURBER, NORWICH, C-T on the left of the hammer.

(I have not attempted a meticulous examination of slight variations in punctuation and diacritical marks in stampings, and I feel they may be disregarded.)

An example of a Dragoon of the Norwich era is shown in Figure 23. This is larger than the Norwich just described but otherwise is not markedly different. The tension spring is now fastened to the upper part of the backstrap, and the plate screw is at the rear. On the two Allens previously described the plate screw was at the front. On the remaining top-hammer, conventional-trigger Allens illustrated in this volume, all made later than this one, the plate screw is invariably located at the rear. Also, on this and on all the remaining top-hammer Allens, the location of the stamping "Allen's Patent" will be found nowhere but on the hammer. On this particular Norwich piece ALLEN & THURBER, NORWICH, C-T is on the left of the hammer, and PATENTED 1837, CAST-STEEL is on barrel ribs.

Though none of the Worcester Allens pictured has grip ornaments, Worcester Allens were made with these ornaments. Also,

Worcester Allens were made with side-plate screws located at the front. It is obvious than an opinion as to when a piece was made at a particular factory can rarely amount to final decision.

In Figure 24 a Dragoon of the Worcester period is shown. This is marked ALLEN & THURBER—WORCESTER and PATENTED —1837—CAST STEEL on the barrel ribs. It has a spur trigger guard, and engraved—instead of stamped—nipple shield ornamentation. It has no silver ovals in the grips.

The Allen shown in Figure 25 is the pocket size and style most frequently seen. The tension screw is now fastened at the bottom and the silver ovals have been discontinued. I have examined several others that appeared to be identical except in the style of engraving and in barrel lengths. One, #7 on all parts, had a 4-inch barrel. Another, #182, had a $4\frac{1}{8}$ inch barrel. The markings on all are the same as on the Dragoon just described.

An Allen, Figure 26, marked ALLEN & THURBER WORCESTER on the barrel and ALLEN'S PATENT 1845 on the hammer, differs from the one pictured in Figure 25 only in having a plain fluted cylinder. The retrogression is under way.

Figure 27 is marked ALLEN THURBER & CO. WORCESTER and PATENTED 1845 on the barrel. This shows further steps in making the guns less costly. The engraving is very meager, and the nipple shield is simply an integral extension of the frame. Finally, this piece is five-shot. All heretofore described are six-shot. Another, not illustrated, bearing #150, is identical except that it is marked ALLEN & THURBER and has the rounded top hammer found on later Allen & Wheelock pieces, instead of the flat top hammer on the Allen Thurber & Co. piece. It provides an example of a supposedly earlier gun made with a later construction.

An Allen pepperbox showing the last economy measures except elimination of engraving is pictured in Figure 28. This is marked ALLEN & WHEELOCK and PATENTED 1845 on the barrel, and ALLEN'S PATENT, JAN. 13, 1857 on the hammer. A major change in construction is the method of fastening the cylinder to the frame. The cylinder is held by a freely turning screw going from the frame into the breech end of the cylinder. This construction is one of the 1857 patent claims. The elimination of the spindle permits a given number of barrels of chosen caliber in a cylinder of less cross-section. A very minor refinement not possessed by any Allen previously described is that the side plate is pinned as well as screwed to the frame.

The last of the top-hammer, conventional-trigger Allens to be illustrated is shown in Figure 29. This five-shot has British proof marks and is stamped on the barrel ALLEN & WHEELOCK. It has no other marks. This is perhaps one of the last and most cheapened of the Allens. It has no ornaments, no barrel ribs, no shield, and, finally, no engraving.

It should not be inferred than an Allen & Wheelock pepperbox with no engraving on the frame is necessarily four- or 5-shot or that it lacks a nipple shield. A pepperbox numbered 733, not illustrated, is marked ALLEN & WHEELOCK on the barrel and PATENTED APRIL 16, 1845 on the hammer. It is six-shot with a fluted cylinder, a screwed-on shield, and no engraving on the frame.

The top-hammer Allen pepperbox, Figure 30, has a ring trigger. It is marked ALLEN & THURBER NORWICH C-T and PATENTED 1837 CAST STEEL on the barrel. It has all the ornamental refinements of the Norwich period and other evidences of being early Norwich, but its grips are not of the shape we expect to find on the early Allens.

The vast majority of Allens were the top-hammers, most of them with conventional triggers; but small numbers of notably different Allens were manufactured.

Illustrations of two Allen pepperboxes, commonly called hammerless, are Figures 31 and 32. To the student and to the collector, the chief difference is that the one with a ring trigger is considerably rarer. Neither is actually hammerless. There is a concealed hammer that fires the top barrel. The frame and the double-action mechanism are modifications of the usual top-hammer assemblies. The nipples are in line with the bores instead of at right angles. Markings are ALLEN & THURBER WORCESTER and PATENTED 1837 CAST STEEL on barrel ribs. There

Fig. 36 Lane & Read—6½" overall—six-shot—.31 caliber—serial #32.

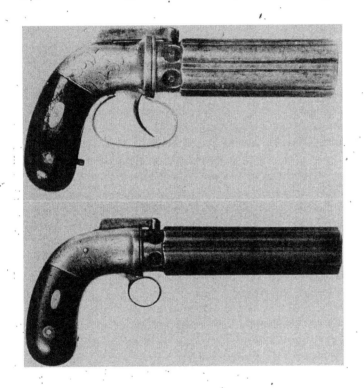

Fig. 37 Tryon—9¼" overall—six-shot—.36 caliber—serial #4.

are minor differences internally in these two concealed-hammer rarities. The outstanding difference is in the use of a straight mainspring in the hammerless model, and of a loop mainspring in the hammer model.

An examination of the scarce model, Figure 33, brings several surprises. It is single-action but quite unlike the original combination single- and double-action model of the 1845 patent in the single-action feature. In the single-action operation of the 1845 version the rotating of the cylinder was effected very suddenly by the pulling of the trigger after the hammer was independently cocked. (There is about an even chance that a late double-action Allen selected at random will permit the hammer's being brought to rest at full cock, without touching the trigger or moving the cylinder.) This single-action Allen has a complicated mechanism which revolves the cylinder by cocking the hammer. The operation is entirely unlike that of single-action pepperboxes of other makes. The stamped markings are ALLEN & THURBER WORCESTER and CAST STEEL on the barrel ribs. There is neither patent date nor "Allen's Patent." The plate screw enters from the right, contrary to custom. The screw is necessary not only to hold the plate; it is essential to the turning of the cylinder.

Established manufacturers sometimes make special brand products for distributors. Allen was no exception. There are many pepperboxes of Allen design made by Allen for wholesalers and dealers. These bear the distributor's name—rarely the Allen firm name—though nearly always they are stamped "Allen's Patent."

The full roll of such distributors is not known. I have examined pepperboxes marked "Bolen," "Lane & Read," "Spies," and "Tryon" which are identical with regular Allens and may be presumed to have been made in their entirey by Allen. Others, marked, "J. Eaton," "Warren & Steele" and Young & Smith" have also been accepted as Allens. Allen probably made pepperboxes for still other dealers. Such pepperboxes will not deviate from the Allen construction, as do those merely externally resembling Allens which are discussed in the next chapter.

A pepperbox made by Allen for J. G. Bolen is shown in Figure 34. This has the same workmanship and all the refinements found on pepperboxes bearing the Allen company name and address. This has the small slotted rear sight that is only occasionally found. It is marked on the barrel PATENTED 1837 CAST STEEL and on the hammer ALLEN'S PATENT and J. G. BOLEN, N. Y. In the 1850's, J. G. Bolen, A. W. Spies & Co., and Young, Smith & Co. were all within a stone's throw of one another in downtown New York. (It seems Allen didn't give exclusive rights and territories.)

A large Dragoon pepperbox marked like the Bolen, except that the name is A. W. SPIES is illustrated in Figure 35. This is of early manufacture, with straight mainspring and fluted ribs, but not of the earliest. It has the pitman and the grip pins which are not found on the earliest Allens.

Another early contract Allen is pictured in Figure 36. This is stamped with the name LANE & READ, BOSTON. Otherwise the markings are like those on the Bolen and on the Spies. George Lane and William Read were in the hardware business at 6 Market Square, Boston, at least as early as 1840. As a pepperbox with their name is seldom seen, they probably did not order many. .

The so-called Dragoon model with the ring trigger, Figure 37, is marked TRYON and CAST STEEL on the barrel. The hammer is marked ALLEN'S PATENT on top and ALLEN & THURBER, NORWICH C-T on the left side. This is probably one of the earliest Norwich pieces. The barrel ribs are fluted and the mainspring is of the early type. It does not have the patent date, which the law passed in 1842 required, and is assumed to be earlier than the Bolen, Spies, and Lane & Read pieces which do bear the patent date.

An impression prevails that Allen pepperboxes were cheap guns a hundred years ago. Still, fifteen dollars for a pocket gun was a considerable outlay at a time when ten cents an hour was a fair wage.

Chapter 5
OTHER AMERICAN
PERCUSSION PEPPERBOXES

THE AMERICAN pepperboxes most frequently found are the Allens. Next are the Blunt & Syms.

The firm of Blunt & Syms did considerable business in firearms imported from Europe. Blunt & Syms pepperboxes have had their provenance questioned because they have European characteristics and often bear no maker's mark. They have a mechanism similar to that of the Mariette pepperbox. Their barrels usually have fluted ribs and their grips, when of wood, have screw plates with studs to prevent turning. These niceties are usual on European pepperboxes; they are uncommon but not unknown on two or three American makes. Orison Blunt in 1849 obtained U. S. Patent No. 6966 for a double-action lock with ring trigger, but this lock was quite different from the familiar underhammer mechanism found on Blunt & Sims pepperboxes which were made for several years before 1849. All this simply means that if Blunt & Syms pepperboxes were not made by Blunt & Syms, they might have been made in Europe.

There is good evidence that Blunt & Syms did make their own pepperboxes. More than one hundred years ago Orison Blunt and John George Syms were manufacturing guns in New York City, and Orison Blunt had been selected by Samuel Colt to make a revolver model. The firm of Blunt & Syms is listed in New York directories all through the 1840's as being manufacturers as well as importers. On the last page of advertisements in Doggett's New York City and Co-Partnership Directory for 1843-4 is an illustration of an underhammer pepperbox described as manufactured by Blunt & Syms. Finally, there is the strong evidence supplied by the broadside pictured in Figure 38. This

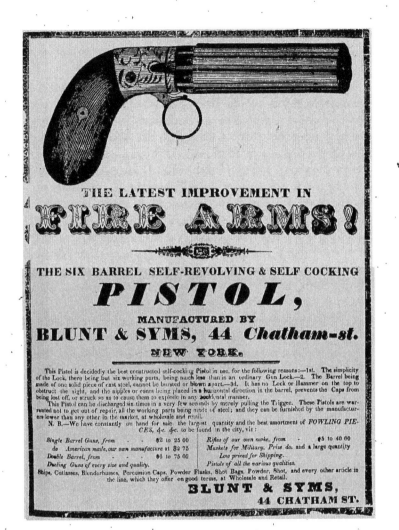

Fig. 38 Blunt & Syms broadside.
(BELLE C. LANDAUER COLLECTION IN THE NEW YORK HISTORICAL SOCIETY.)

refers to the pepperbox as being manufactured —not "sold" or "imported"—by Blunt & Syms.

The pepperbox pictured in Figure 39 is marked O. BLUNT PATENT 1849 and is classed here as Blunt and Syms. It is the rarest and most interesting in the Blunt & Syms division. This is the only example known to me that has the lock construction just as it was patented by Mr. Blunt. This piece has the double-action top-hammer and ring-trigger construction designed by Mr. Blunt for use on a single-shot pistol. The trigger is not designed to spring forward when the hammer falls, but must be pushed forward by the finger between shots. This construction may be adequate in a single-shot pistol, but it seems unsuited to a multi-shot firearm. The cylinder is mechanically turned by the usual trigger-operated ratchet. Mr. Syms was not a joint patentee with Mr. Blunt, and it may be he was not associated with Mr. Blunt in the manufacture of this rarity.

In the usual top-hammer pepperbox the nipples are at right angles to the cylinder, and the barrel in firing position is the one at the top. In the the concealed-hammer Blunt & Syms the nipples are in line with the barrels, and the barrel in firing position is the one at the bottom. This in-line construction of the nipples is more compact and may give better ignition than the right-angle construction on top-hammer pepperboxes. The ring trigger, when forward, engages a clutch in a hammer slot; when drawn fully back, the clutch slips off and the hammer is driven forward by the mainspring. Trigger pull also causes the cylinder to rotate. This type of action was very popular in Continental Europe and moderately popular in England. On the Continent individual unscrew barrels were favored; in England solid cylinders like the Blunt & Syms were preferred.

The heavy trigger pull of these guns was no aid to accuracy. A pepperbox trigger with a light pull means a weak mainspring which may not drive the hammer with sufficient force to explode the cap. Sometimes especially sensitive caps were used on pepperboxes, but their use entailed greater danger of all barrels being fired at once.

Figure 40 is of an unusually small Blunt & Syms. It has

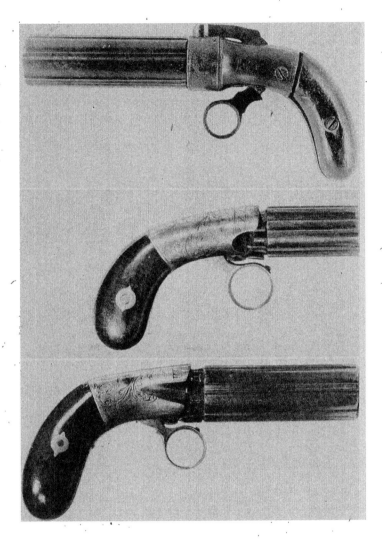

Fig. 39 (*top*) O. Blunt—8¼" overall—six-shot—.31 caliber—serial #89.
(SAM E. SMITH COLLECTION.)

Fig. 40 (*center*) Blunt & Syms—5" overall—five-shot—.24 caliber.
(ANTHONY FIDD, JR., COLLECTION.)

Fig. 41 (*bottom*) Blunt & Syms—7½" overall—six-shot—.30 caliber.

the usual blued iron barrels, engraved iron frame, and smooth walnut grips ordinarily found on Blunt & Syms and most other American pepperboxes. Figure 41 shows a Blunt & Syms of the size most frequently seen. A Blunt & Syms pepperbox, unusual in that it has a smooth round barrel, is pictured in Figure 42.

The Blunt & Syms, Figure 43, has a saw-handle grip that did not find favor with the other American makers, except for some single-shot pistols. The other saw-handle Blunt & Syms, Figure 44, is uncommonly large. It is noticeably different not only in size but in shape of frame and grips.

An exceptional Blunt & Syms is pictured in Figure 45. It has engraved silver grips and a deeply fluted cylinder.

Considerable lack of uniformity will be noticed in the sizes and shapes of cap slots cut in Blunt & Syms frames, and in the extent to which the nipples are covered by the frames. In the latter regard, there may have been divided opinion whether to protect the nipples better with a large shield or to decrease risk of multiple explosions with a small shield. Some Blunt & Syms pepperboxes have no shields whatever.

While some American pepperboxes, like the Blunt & Syms, are markedly different from Allens, others strongly resemble Allens. In the preceding chapter there was mention of the Spies, Bolen, and other pepperboxes made for various dealers by Allen. These not only look like Allens—they are Allens. Other pepperboxes that look more or less like Allens were made by rivals and bear such names as Marston, Manhattan Arms Co., Washington Arms Co., Union Arms Co. These have sometimes been considered infringements on Allen patents, but a careful examination, and study of the Allen patents shows that the assumption was false. The mechanism patented by Allen in 1845 was never copied. The 1837 Allen patent, covering only a double-action lock, ran only fourteen years and expired in 1851. The evidence is that these competitive pepperboxes were not manufactured until the 1837 patent expired.

Manhattan pepperboxes have been variously described as made by Allen and as infringements. Superficially some Manhattans do look remarkably like some Allens. Omission on Man-

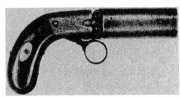

Fig. 42 Blunt & Syms—6¾" overall—six-shot.
(WILLIAM M. LOCKE COLLECTION.)

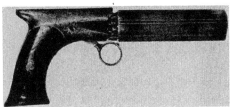

Fig. 43 Blunt & Syms—9" overall—six-shot—.31 caliber.

Fig. 44 Blunt & Syms—10¼" overall—six-shot—.38 caliber.

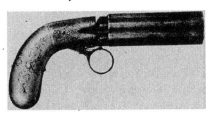

Fig. 45 Blunt & Syms—7¼" overall—six-shot—.31 caliber.

hattans of "Allen's Patent" found stamped on pepperboxes made for Allen contract dealers gives circumstantial evidence, and examination of construction gives direct evidence that they were neither infringements nor of Allen manufacture.

Figure 46 is a double-action Manhattan, with mechanically rotated cylinder, that looks like an Allen. Disassembling this Manhattan reveals that the joining of trigger to hammer is quite unlike the assembly on the Allen. The mainspring is unlike that of the Allen, and it is fitted and operated differently. The center pin on this Manhattan screws into the frame, and removal of the cylinder requires taking out the center pin. This five-shot pepperbox has MANHATTAN F. A. MFG. CO. NEW YORK stamped on the left of the hammer. Another Manhattan, not illustrated, a six-shot numbered 40, has the center pin solidly fixed in the frame, and it has the stamped name on the cylinder. When the small holding screw at the muzzle end of the center pin is taken out, the cylinder will slide off the pin.

Figure 47 shows a double-action Manhattan with a hand-revolved cylinder. The barrels are turned manually against the pressure of a small spring-backed plunger, set in the frame, which engages a slot in the cylinder when a nipple is under the hammer. The barrel is held by a screw through the frame, inserted vertically down to the cylinder axis.

Pepperboxes marked "Union Arms Co." and pepperboxes marked "Washington Arms Co." are identical in appearance and construction. It would seem they were all made by the same manufacturer. It is not definitely established that the firm named Union Arms Company, located in Hartford, Connecticut, in 1861, made them. No address for Washington Arms Co. has been found. One of these pepperboxes—this happens to be marked WASHINGTON ARMS CO.—is illustrated in Figure 48. Except that this has a shield riveted instead of screwed to the frame, it looks like an Allen externally. Internally it is quite different, of simpler and perhaps more effective construction. The mainspring is fastened to the opposite portion of the backstrap and bears directly on the hammer. The double-action is effected by a pawl hinged to the trigger, driving upward on the breast of the hammer.

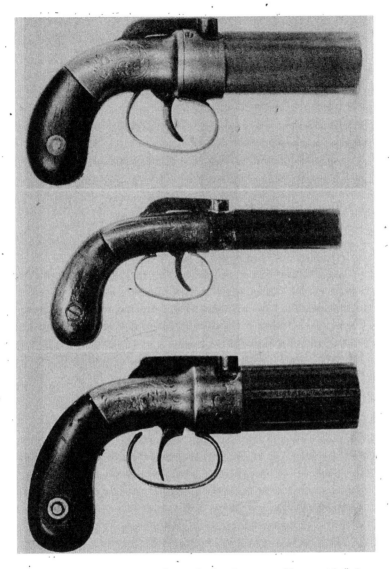

Fig. 46 (*top*) Manhattan—5¾" overall—five-shot—.30 caliber—serial #482.
Fig. 47 (*center*) Manhattan—6" overall—three-shot—.30 caliber—serial #62.
Fig. 48 (*bottom*) Washington Arms Co.—6½" overall—six-shot—.32 caliber—serial #27.

54 OTHER AMERICAN PERCUSSION PEPPERBOXES

Two pepperboxes that resemble Allens in all but the lock mechanisms are the Bacon and the Stocking. These could have been, and probably were, legally made before expiration of the Allen 1837 patent. Both are single-action and cannot be fired as quickly as a double-action Allen, but in their day some men preferred single-action pepperboxes just as today some men like the old single-action revolvers.

Figure 49 pictures a single-action underhammer Bacon. The gun is cocked and the cylinder rotated by drawing back the underhammer by its trigger-like projection. It is fired by pressing the rear trigger. An advantage of the single-action is better aim, but this gun is not provided with sights. The Bacon had no top hammer to distract aim, and on no other pepperbox would sights have been more practical. This Bacon, marked on the barrel BACON & CO. NORWICH C-T and CAST STEEL, has a ribbed barrel. Others have barrels without ribs.

Incidentally, the individual characteristics of the stamping "Norwich C-T" on all the Bacons I have examined under a glass appear identical with those found on some Norwich Allens. Markings by different stamps are as individual as fingerprints. Allen used more than one "Norwich C-T" stamp. Perhaps Bacon used a discarded Allen stamp.

Of the regularly manufactured pepperboxes, Bacon is the only single-action underhammer. It and one model Pecare & Smith are probably the only ones, after the Darling, with a cylinder revolving clockwise. Mr. Thomas K. Bacon did one other thing with his pepperbox that was almost unheard of among manufacturers of percussion-cap pepperboxes. He ran an advertisement picturing it. Figure 50 is a reproduction of the advertisement that appeared in an early Connecticut directory.

The Stocking, Figure 51, is of about the same size and construction as the Bacon, with the important difference that it is a top-hammer. Pressure on the hammer's long projecting end cocks it and rotates the cylinder. The trigger is conventional, with a guard. The gun is marked on the ribbed barrel STOCKING & CO. WORCESTER and CAST-STEEL WARRANTED.

Figure 52 shows another Stocking with spur trigger guard

Fig. 49 Bacon—7½" overall—six-shot—30 caliber—serial #10.

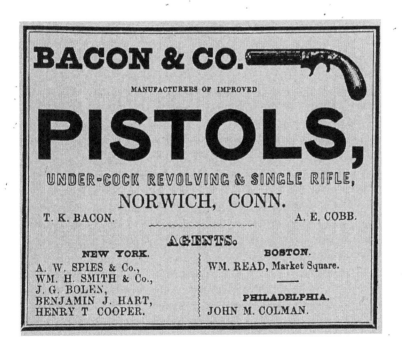

Fig. 50 Bacon advertisement in Connecticut Directory.
(COURTESY JOHN HINTLIAN)

and with oval rather than diamond silver ornaments in the grips.

A Stocking pepperbox that is unusual by reason of its long barrel is shown in Figure 53.

Most Stocking pepperboxes have nipple shields and nipples integral with the cylinder. The Stocking, Figure 54, is probably an early model. It was made without nipple shield and with removable nipples.

The pepperbox that turns up—very infrequently—at a country auction or elsewhere is not likely to be in fine condition; but if it be a Stocking, it will very probably be in working order.

Mr. W. W. Marston had a place of business at Second Avenue and 22nd Street, New York, which was the address of the Phoenix Armory. Marston pepperboxes which follow the Allen lines but not the internal construction are found with varied markings. Markings seen or reported on Marston firearms include: W. W. Marston, New York; W. W. Marston, New York City; W. W. Marston Armory, New York; Wm. Marston Phoenix, New York City; Phenix (sic) Armory; Marston & Knox; W. W. Marston & Knox, New York. It should be noticed these are the William Walker Marston guns, different from the Stanhope W. Marston pepperboxes to be described later.

A pepperbox marked PHENIX ARMORY, Figure 55, has the Allen look. It has a riveted shield, like that of the Washington Arms Co. pepperbox. Internally the mechanism is unlike the Allen, particularly in the direct connection of trigger and hammer and the fitting and shape of the mainspring. This has a fluted cylinder.

A pepperbox marked W. W. MARSTON & KNOX, NEW YORK is pictured in Figure 56. It is like the Phenix Marston except that it has a ribbed barrel and a wider shield.

A pepperbox, marked SPRAGUE & MARSTON, is shown in Figure 57. The engraving of the eagle head on the frame is unusual, as is the use of a non-turning plate for the grip screw. It is generally believed the Marston of Sprague & Marston was W. W. Marston, and that Sprague & Marston pepperboxes are later than those marked with only the W. W. Marston name. However, the listing of Sprague & Marston as "Gun and Pistol

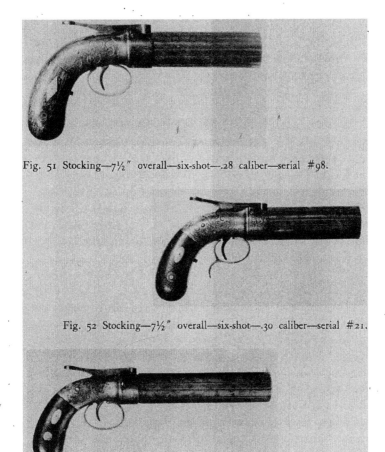

Fig. 51 Stocking—7½″ overall—six-shot—.28 caliber—serial #98.

Fig. 52 Stocking—7½″ overall—six-shot—.30 caliber—serial #21.

Fig. 53 Stocking—9½″ overall—six-shot—.30 caliber.
(FRANK R. HORNER COLLECTION.)

Fig. 54 Stocking—7½″ overall—six-shot. (WILLIAM M. LOCKE COLLECTION.)

Makers" in the New York business directory of 1850-51, with the address "Jane corner Washington," indicates Sprague & Marston may have been earlier.

Another Sprague & Marston, unusual in that the barrel is round, is shown in Figure 58.

On September 18, 1855, Mr. W. W. Marston obtained Patent No. 13581 for a pepperbox with several innovations. Instead of the center pin's being attached to the frame and entering the cylinder, it was attached to the cylinder and entered the frame. The end of this spindle in the frame was provided with cams, against which the trigger operated. Firing required extraordinary procedure: the trigger was pulled to partly rotate the cylinder and bring the hammer to half-cock; the trigger was released and pulled a second time to nearly complete rotation of the cylinder and bring the hammer to full cock; the trigger was released and pulled a third time to complete rotation of the cylinder and drop the hammer. Mr. Marston used thirty-three letters and numbers on his patent drawing to identify parts and explain operation. He may have had misgivings; he explained the gun could be constructed so that one of the three trigger pulls would be eliminated. Figure 59 is of the Smithsonian Institution's patent model. It is doubtful that any of these pepperboxes was marketed.

The pepperboxes described in the remainder of this chapter are all unlike top-hammer Allens in appearance as well as mechanism. Probably the most interesting of these to collectors is the ten-shot Pecare & Smith.

Jacob Pecare and Josiah Smith, of New York, obtained Patent No. 6925 on December 4, 1849. Surprisingly, their patent claim was for nothing but the concealed trigger. The unique hammer and other features that would appear to have been patentable were not claimed, though described in the specifications and shown in the patent drawing.

The mechanically revolved cylinder of a 10-shot double-action Pecare & Smith is attractively ribbed, but it is completely hidden by the fixed shield inside which it revolves. The reason the shield was made the full length of the cylinder, instead of

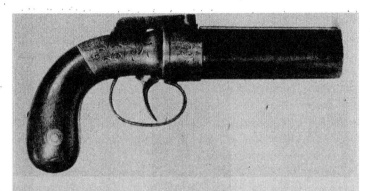

Fig. 55 Phenix Armory (*sic*)—7½" overall—six-shot—.32 caliber—serial #4.

Fig. 56 Marston & Knox—7½" overall—six-shot—.31 caliber—serial #24.

Fig. 57 Sprague & Marston—6½" overall—six-shot—.28 caliber.
(DR. PRENTISS D. CHENEY COLLECTION.)

just long enough to cover the nipples, is explained in the specifications which read in part: "The nature of our invention consists in so constructing a (pepperbox) that an assailed party cannot by grasping the barrels prevent the assailant from repeating his fire; and that the concealment of the trigger, hitherto unattained in revolvers, but particularly desirable in this kind of pistol is effected." The concealed trigger is jointed and folding; it opens when pressed back.

The ten-shot Pecare & Smith is sometimes called hammerless, but actually the hammer is not even concealed unless one looks from the side or from underneath. The small hammer is placed vertically in, and below the top of, the frame. The frame and shield are cut away at the top for the length of travel of the hammer. Pressure on the trigger draws the hammer back and slightly upward. Just before final pressure drops the hammer, a groove in the rising nose of the hammer provides a rear sight. There is a front sight on the shield.

The ten-shot Pecare & Smith illustrated, Figure 60, has a brass frame and iron shield, both engraved. The grips are of walnut. The cylinder sleeve is marked PECARE & SMITH PATENT 1849.

The four-shot Pecare & Smith is perhaps rarer than the ten-shot, though either is more frequently found than some other American pepperboxes illustrated here. Both were made with iron frames as well as brass. An example of a four-shot Pecare & Smith is shown in Figure 61. The patented trigger is in evidence, but otherwise it is dissimilar. The barrels, with individual sights, are turned manually and are not covered with a shield. The reason the uncovered cylinder has only shallow fluting while the covered cylinder of the ten-shot is finely ribbed, is a perplexity to the collector. The hammer of this four-shot is double-action, but of the usual bar type. This piece has just the "Pecare & Smith" marking.

The two Pecare & Smith pepperboxes just described are both double-action, but one has a mechanically turned and the other a manually turned cylinder. The four-shot thumb-hammer Pecare & Smith pictured in Figure 62 is single-action with a cylinder

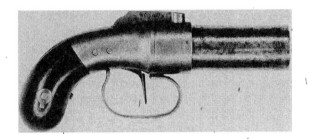

Fig. 58 Sprague & Marston—6½" overall—six-shot.
(WILLIAM M. LOCKE COLLECTION.)

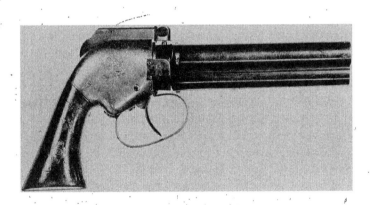

Fig. 59 W. W. Marston Patent Model—9" overall—three-shot—.40 caliber.
(SMITHSONIAN INSTITUTION COLLECTION.)

which is mechanically revolved by means of a ratchet and a pawl linked to the hammer. The trigger is not of the patented type but of the conventional concealed type that flips open when the hammer is cocked. Only one other combination of lock action and cylinder turning is possible—a single-action lock with a manually turned cylinder. Such a Pecare & Smith is pictured in Part II of the *Bulletin of the Public Museum of the City of Milwaukee*, dated May 25, 1928, Plate 86, Figure 23. The illustration shows a thumb hammer and a conventional trigger with guard.

Four-shot Pecare & Smith pepperboxes with ring triggers have been reported, but none is available for illustration. Also, four-shot Pecare & Smith pepperboxes with patented triggers, double-action locks and mechanically revolved cylinders are known to exist. Differences noticed between one of these last-mentioned four-shots (one which looks much like Figure 61) and the ten-shot illustrated, include: the four-shot cylinder revolves counter-clockwise but the ten-shot revolves clockwise; the four-shot cylinder is fastened directly by a screw from inside the frame but the ten-shot cylinder is held in place by a large screw fastened in the muzzle end of the long center pin; the four-shot has vertical nipples but the ten-shot has horizontal nipples. The hammers, the mainsprings, and the mechanisms that turn the cylinders and raise the hammers are designed differently in the two sizes.

According to the Official Catalogue of the London 1851 Exhibition, Pecare & Smith exhibited "repeating pistols with stocks of ivory and rosewood, mounted with steel and gold." (At the 1851 Exhibition there were 138 manufacturers including five from the United States, displaying firearms. As this is written, one hundred years later, there are only eight manufacturers of guns exhibiting at the 1951 British Festival, the displays being only those pieces of top quality—British "Best" guns.)

The very fine matched pair of engraved ten-shot Pecare & Smith pepperboxes, gold trimmed with ivory grips, shown in Figure 63, may possibly have been displayed at the 1851 Exhibition. These have the cylinder sleeves marked PECARE & SMITH FIREARMS MANUFACTORY AMERICAN STEAM WORKS 180 & 182 CENTER STREET NEW YORK.

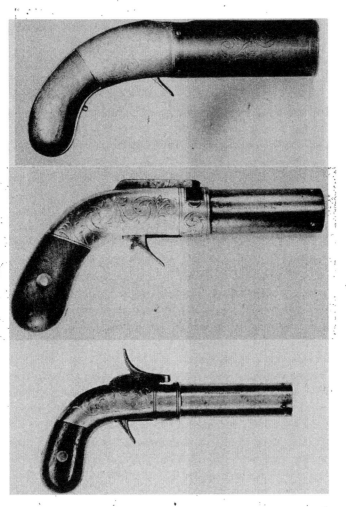

Fig. 60 (*top*) Pecare & Smith—8¼" overall—ten-shot—.28 caliber—serial #4.
(R. H. MC LEOD COLLECTION)

Fig. 61 (*center*) Pecare & Smith—6½" overall—four-shot—.31 caliber.
(DR. PRENTISS D. CHENEY COLLECTION.)

Fig. 62 Pecare & Smith—7" overall—four-shot.
(WILLIAM M. LOCKE COLLECTION.)

OTHER AMERICAN PERCUSSION PEPPERBOXES

In the gold rush days, several men who worked for Allen & Thurber left that company and either made their own pepperboxes or worked for other makers. Alexander Stocking, who made the Stocking pepperbox, was one. Another was George Leonard, Jr., of Shrewsbury, Massachusetts. Mr. Leonard's Patent No. 6723, of September 18, 1849, covered a striker that revolved and a cylinder that did not revolve. His later Patent No. 7493, granted July 9, 1850, embodied the other peculiarities found in Leonard pepperboxes.

Figure 64 is of a four-shot Leonard marked on the barrel G. LEONARD JR. CHARLESTOWN PATENTED 1849 CAST STEEL, but it has the distinctive features of the 1850 patent. The unusual characteristics are: the cylinder that is screwed into the frame and does not revolve; the striker that revolves; the two triggers. The patent specifications direct that to fire the pistol, the ring trigger be first pulled back by the middle finger, and then the forward trigger be pulled by the forefinger.

The 1850 patent drawing shows a center barrel, rifled, to be fired when desired by pressing the hammer to position to strike a central nipple. It is doubtful that more than a very few were made with this complication.

Very few percussion-cap pepperboxes were made with rifled barrels. The Leonards, except for the one just mentioned, were probably all smooth bore. However, the Robbins & Lawrence developments of the Leonard do have rifled barrels.

It is interesting to note that the 1850 patent specifications made reference to the barrels being "drilled and bored entirely through," but did not claim the invention. When the Rollin White patent covered this a few years later, the result was far-reaching, giving Smith & Wesson a virtual monopoly on cartridge revolvers for years.

Robbins & Lawrence, of Windsor, Vermont, made pepperboxes under the Leonard patent and improved the construction. The pepperbox in Figure 65 is marked ROBBINS & LAWRENCE CO. WINDSOR VT. PATENT 1849 on the barrel. The cylinder is in two parts to facilitate loading. The forward part is bored through. The rear part contains the powder chamber and the

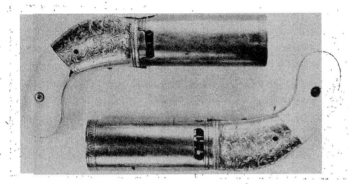

Fig. 63 Pecare & Smith matched pair—8" overall—ten-shot—.28 caliber.
(DR. PRENTISS D. CHENEY COLLECTION.)

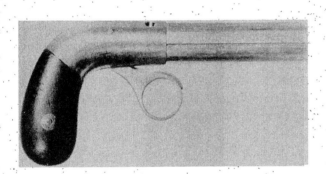

Fig. 64 Leonard—6½" overall—four-shot—.31 caliber—serial #72.

nipples; it is hinged at the bottom and drops down for capping when a catch on top of the frame is pressed. This change from the screwed-on cylinder of the Leonard makes for quicker recapping and also assures the cap's lying under the hammer. The internal construction is much sturdier, and the mechanism for revolving the hammer is entirely changed. In the Leonard the hammer is fixed to a round plate which is held loosely in a groove and revolved by a ratchet. In the Robbins & Lawrence the rotation is accomplished by means of inclined grooves in the striker. The button on the backstrap is simply to permit easing forward the ring trigger when one wishes to let down the hammer without firing. When this button is defective, as it frequently is, pressure on it will fire a cocked gun.

A larger Robbins & Lawrence is illustrated in Figure 66. The rear of the cylinder is shaped differently; the cylinder-release catch is shorter; and the mechanism is slightly modified. On both these Robbins & Lawrence pepperboxes the front sight is midway between the top barrels. Another of these larger pepperboxes, Figure 67, has the cylinder fitted so the front sight is directly over the barrel. This change makes the muzzle appearance conform to that of other makes but does not improve aiming or shooting because on these guns it is the striker, not the cylinder, that revolves. In either barrel grouping of these five-shot pepperboxes, the barrel being fired may be directly under, or it may be at the right or left of the line of sight. These two larger pieces have other very minor differences in hinge construction and cylinder shape.

Figure 68 pictures a group of Robbins & Lawrence pepperboxes from the William M. Locke collection. Four of the small size with the usual ring-trigger five-shot construction are shown, two on each side at the bottom. Of these, the one at the extreme right is profusely engraved and fitted with ivory grips. The one next to it, which has no engraving, is similar in two other respects to the unengraved outsize piece in the center. Both have brass frames and both are unmarked. Both pieces, unquestionably made by Robbins & Lawrence, appear to be experimental models. The very large one, 11½ inches overall, which

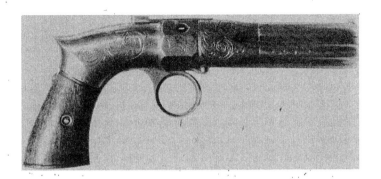

Fig. 65 Robbins & Lawrence—7½" overall—five-shot—.28 caliber—serial #4130.

Fig. 66 Robbins & Lawrence—9" overall—five-shot—.31 caliber—serial #5611.

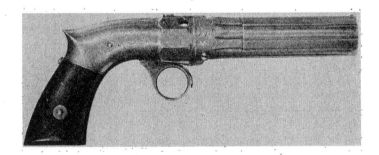

Fig. 67 Robbins & Lawrence—9" overall—five-shot—.31 caliber—serial #4245.

is six-shot, almost surely was never put in production, while the small one was probably never regularly manufactured with a brass frame. Just below the largest Robbins & Lawrence in the center is another unmarked Robbins & Lawrence, four-shot, with a conventional trigger. This is probably unique. Knowing of the association of Robbins & Lawrence with Christian Sharps, as well as with George Leonard, and keeping in mind that Christian Sharps patented a percussion-cap pepperbox with a revolving inside striker and a conventional trigger, one wonders if Robbins & Lawrence experimented to combine the ideas of Leonard and Sharps and to supplement their inventions with a double-action mechanism. The three pepperboxes at the top of the plate are the usual large size. The one at the upper left has an underneath set screw to insure alignment of breech and barrels; the one at the upper right has four rectangular vents leading to the nipple cavity in addition to the two circular vents that are standard construction.

The pepperbox in Figure 69 appears from its frail construction to be an experimental model, hardly designed for use. On the backstrap is engraved J. P. TIRRELL MAKER NORTH BRIDGEWATER. The mechanically rotated cylinder is of smooth bronze. The placing of the mechanism is unlike anything ever seen on another pepperbox. The mechanism is all fastened to the flat frame between the grips. What seems to be the backstrap is just the edge of this solid bronze frame. The left grip is hollowed to a thin shell to give room for the intricate and delicate works. This Tirrell is somewhat like the Blunt & Syms in its concealed hammer and ring-trigger double-action construction.

The Post pepperbox, Figure 70, is the Smithsonian Institution's patent model. Jacob Post's patent of May 15, 1849, No. 6453, covers only a device for setting a trigger on a double-action gun, "of any approved fashion." To fire a pepperbox with Mr. Post's invention requires first pulling, then releasing the trigger. The hammer is not released until the ring trigger moves forward. The screw under the frame, just back of the trigger, adjusts the hammer release. Mr. Post contrived some-

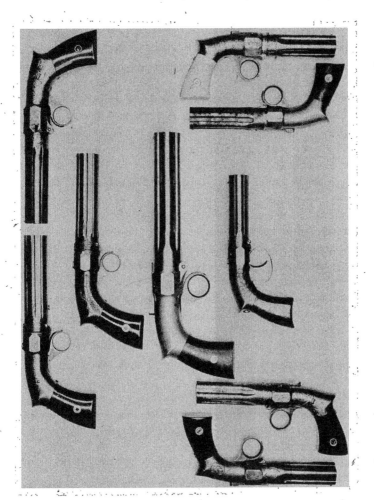

Fig. 68 Group of Robbins & Lawrence pepperboxes. (WILLIAM M. LOCKE COLLECTION.)

thing new, but his inspiration was evidently considered of little worth. The patent model is marked J. POST NEWARK N. J. Those commercially produced were marked J. POST'S SELF ACTING SET PATENTED 1849. Few are found.

Another pepperbox that probably had even less sucess was the Chamberlain with its double-action ring trigger and upright hammer, patented April 23, 1850, Patent No. 7300. Figure 71 is of the patent model in the Smithsonian Institution. Mr. Chamberlain was granted his one claim: "The improved mode of attaching the cylinder to the stock . . . whereby I dispense with the usual spindle and hole for its reception in the center of the cylinder of barrels, being thus enabled to enlarge the bore of the barrels in a cylinder of equal size." In the specifications Mr. Chamberlain stated, "I have sought to make a simple, cheap, and effective weapon, one not liable to get out of repair, and which will answer for ordinary service of travelers, hunters, or backwoodsmen." His pepperbox would seem to accomplish what he sought in construction, but it was not accepted in the market he anticipated. The trouble was that Mr. Chamberlain achieved the elimination of the spindle and the increase in fire-power only by making the cylinder hand turning. There was demand only for pepperboxes with mechanically rotated barrels.

The pepperbox illustrated in Figure 72 is cocked by pulling down the underhammer until it clicks into engagement with the trigger. The iron cylinder is turned by hand against the pressure of an inside spring without the use of a separate release mechanism. The fine workmanship and silver inlay ornamentation, together with the clear E. B. WHITE made with a single stamping on the backstrap, suggests that Mr. White was an established gun maker, but when or where he was in business is unknown.

This E. B. White pepperbox and the unmarked pepperbox pictured in Figure 73 would both be called primitive if the word did not connote crudity. Neither deserves the unfavorable implication. The clues to the maker of an unmarked pepperbox are sometimes too slight and contradictory to warrant more

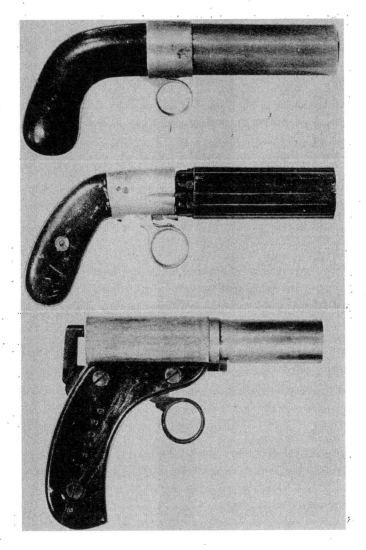

Fig. 69 (*top*) J. P. Tirrell—7″ overall—six-shot—.26 caliber.
(ROBERT ABELS COLLECTION.)

Fig. 70 (*center*) J. Post Patent Model—7½″ overall—six-shot—.30 caliber.
(SMITHSONIAN INSTITUTION COLLECTION.)

Fig. 71 (*bottom*) Chamberlain Patent Model—6″ overall—three-shot—.32 caliber.
(SMITHSONIAN INSTITUTION COLLECTION.)

than a guess, but often an examination of an unmarked pepperbox will yield evidence amounting to proof.. The construction of the single-action hand-turned pepperbox, Figure 73, strongly suggests that the maker was William Billinghurst, the only American maker of his day recognized, here and abroad, as the equal of the best British makers. The underhammer with the trigger guard as a mainspring is typical of Billinghurst bench gun construction, and the peculiar adjustable rear sight is found on Billinghurst "Buggy rifles." The numbered barrels, which have front sights, screw into the breech which contains the nipples.

The pepperbox shown in Figure 74 has the construction of grips, frame, and underhammer that is typical of the single-shot bootleg pistol. It is unmarked and possibly has no counterpart. In operation it is like the first-model Darling in that it is a single-action pepperbox with a mechanically rotated cylinder. The construction of hammer and trigger is seen at a glance to be quite different, but the barrel fluting is like that on a Darling. In Issue No. 21, *The Gun Collector* speculated on the possibility that an individual named Goddard assembled some pistols from Darling parts. Whether Goddard had anything to do with making this pepperbox is unknown.

Another unmarked pepperbox that appears to be American is shown in Figure 75. This is double-action with a mechanically revolved cylinder and is of advanced style both in form and mechanical construction. It has been suggested that this fine pepperbox may have been made by William Billinghurst, but it cannot be confidently attributed to that famous maker.

Figure 76 is an unmarked pepperbox with a mechanism which in a simple and effective way rotates the cylinder as the hammer rises and securely locks the cylinder in position at the instant the hammer falls. Three of its features—the direct action of the trigger on the hammer, the placing of the mainspring, and particularly the necessity after firing of pressing the trigger forward against spring pressure to re-engage the hammer—are points in the patent granted Stanhope W. Marston, January 7, 1851, Patent No. 7887. A careful examination of the gun in

Fig. 72 E. B. White—7½″ overall—six-shot—.28 caliber.
(SAM E. SMITH COLLECTION.)

Fig. 73 Unmarked—10¼‴ overall—six-shot—.25 caliber.
(DR. PRENTISS D. CHENEY COLLECTION.)

Fig. 74 Unmarked—8½″ overall—six-shot—.31 caliber.
(SAM E. SMITH COLLECTION.)

Fig. 75 Unmarked—6¾" overall—six-shot—.36 caliber.
(DR. PRENTISS D. CHENEY COLLECTION.)

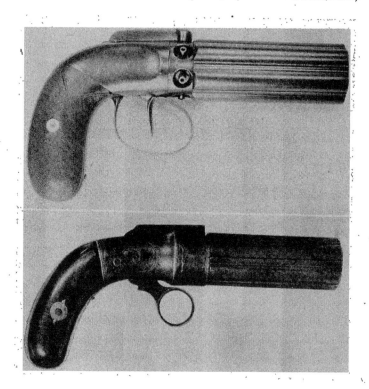

Fig. 76 Unmarked—7½" overall—six-shot—.31 caliber.

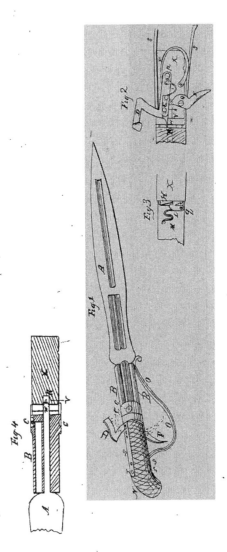

Fig. 77 Lawton Pistol-Saber—Patent Drawing.

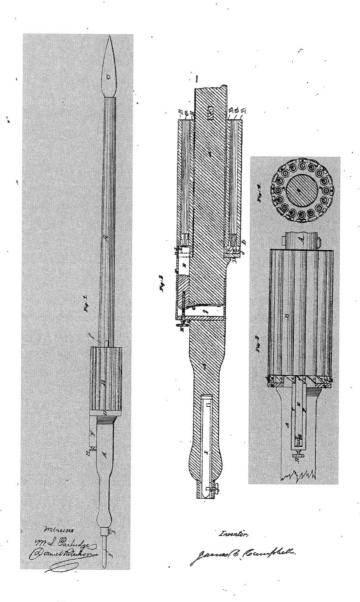

Fig. 78 Campbell Lance-Pepperbox—Patent Drawing.

connection with a study of the patent drawings and specifications leads me to attribute the piece to this native of England who lived in New York in the 1850's. What differences there are between the actual gun and the patent specifications tend to confirm rather than disprove the presumption. The 1851 S. W. Marston patent provided for the rotating of the cylinder by releasing tension on a spring, and it did not provide a cylinder lock. Keeping a spring under continual tension when not in use was hardly a style likely to be followed in production. The simple method used to turn the cylinder in the gun illustrated is an improvement. The addition of a positive cylinder lock is another improvement made without alteration of the action.

The patent drawings of two extraordinary pepperboxes that were probably never marketed commercially are shown in Figures 77 and 78.

Though the patent for the Lawton pistol-saber, Figure 77, was not granted until November 23, 1837, the only known example of this combination weapon has a blade marked N. P. AMES SPRINGFIELD 1835. It is unlikely the invention was fully conceived in 1835. As the drawing shows, the cylinder revolved around the shank of the saber. Drawing back the trigger turned the cylinder to bring a nipple under the hammer, which was simultaneously raised by the trigger pull. To fire the cocked gun, the trigger was pressed forward. To fire another barrel, the pulling back and the pressing forward was repeated.

Mr. J. C. Campbell's patent, Figure 78, was for another military arm which had a sixteen-barrel cylinder revolving on the shaft of a lance. At close quarters, with the lance leveled at an enemy, this could have been a fearsome weapon, capable of very rapid fire if it functioned as planned. "With the cylinder grasped in the left hand and the back part of the pole in the right hand" a one-sixteenth revolution of the cylinder fired a charge. This manual turning resulted in the edges of teeth at the rear of the cylinder pushing back a hammer against spring pressure. It was assumed when one of the teeth cleared the nose of the hammer, the hammer would be driven forward quickly enough and hard enough to hit and detonate the percussion cap. In his specifications Mr. Campbell did not overlook fitting the butt of the lance with a movable spike to drive in the ground as an aid in stacking the arms, nor to fashion the steel lance head in "such form as to cut both in being pushed forward and in being drawn from a wound."

Chapter 6
THE SHARPS PEPPERBOXES

THE POPULARITY of percussion-cap pepperboxes had waned greatly before the invention of the first satisfactory rim-fire cartridge. Hand guns of better accuracy and greater hitting power were wanted, and the demand was for revolvers.

There would probably have been few cartridge pepperboxes if Smith & Wesson's patent on a breech-loading rim-fire revolver had not exercised a virtual monopoly. The Smith & Wesson— originally the Rollin White—patent prevented others from making a revolver with the cylinder bored through, but it did not prevent the manufacture of pistols or pepperboxes with barrels bored through. Accordingly, several manufacturers produced pepperboxes using rim-fire cartridges. Surprisingly, Bacon was the only maker of percussion-cap pepperboxes to put out cartridge pepperboxes. The other cartridge pepperboxes were manufactured for the most part by established makers of percussion-cap arms other than pepperboxes. Some cartridge pepperboxes were made by newcomers.

As Allen was the leader in the percussion-cap field, so was Sharps in the cartridge field. Christian Sharps undoubtedly made cartridge pepperboxes surpassing in number the combined outputs of all other American makers, but his fame rests on the long guns bearing his name. In 1848 he invented a breech-

Fig. 79 C. Sharps Patent Model—9" overall—four-shot—.38 caliber.
(SMITHSONIAN INSTITUTION COLLECTION.)

loading percussion-cap rifle that was developed into the most popular gun of its day, but in 1853 he disassociated himself, even as technical adviser, with the company manufacturing it and gave his attention to pistols. Up to 1859, along with the pistols, he made some pistol-rifles and revolvers, all percussion-cap and none in quantity. In 1859 he obtained a patent for a "breech-loading repeating firearm" and from then on his chief concern was the manufacture of those arms.

It is a little surprising that Christian Sharps' U.S. patent, granted Jan. 25, 1859, makes no mention of two features included in the Provisional Specification of his British patent, which was dated Jan. 22, 1859 and sealed March 15, 1859. A catch fitting loosely on the hammer—designed as a safety feature—and a hard steel center piece in the barrel block to act as an anvil to receive the hammer blow were both claimed as features in the British Specification. As far as I know, neither of these features was included in Sharps pepperboxes, made either in this country by Sharps or by Tipping & Lawden under the Sharps patent in England.

The U.S. patent drawing names the gun a "revolver," though its barrels do not revolve. Some call it a "derringer," though the single shot pistol made famous by Henry Deringer is altogether different. As a generic name, pepperbox seems best.

A sales argument that Christian Sharps and his distributors used was based on the fact that with his guns there was no escape of gas from the breech. He emphasized that in a gun with revolving cylinder and separate barrel there was loss of explosive force due to escape of gas at the joint between cylinder and barrel, and that consequently his firearm gave greater penetration than any other using a like cartridge.

Before proceeding to consideration of the Sharps cartridge pepperboxes, brief attention should be accorded a single-action percussion-cap pepperbox patented by Christian Sharps in 1849, Patent No. 6960. Figure 79 shows the patent model now in Washington. This is probably the only one made. The invention comprised a method of cocking and simultaneously revolving a striker by means of a lever-operated ratchet. Leonard's patent

(Chapter 5) was analogous and slightly earlier. (Robbins & Lawrence were responsible for the great production and sale of Sharps rifles. Whether or not Mr. Sharps approached them about the pepperbox is unknown. At any rate, the Robbins & Lawrence pepperboxes were made under the Leonard patent.)

The Sharps patent of Jan. 25, 1859, #22753, covered a single-action pepperbox with a revolving projection on the hammer which struck the edges of rim-fire cartridges. This 1859 patent probably had its genesis in the 1849 patent. Mr. Sharps stated that only one form of cartridge was to be used. He described a rim-fire cartridge and specified that the barrel bores were to be recessed for the cartridge heads. He also described an extractor of a sort apparently used only on the model with a barrel release in the form of a lever.

Pressure on a spring is the usual Sharps method of releasing the barrels to permit their being slid forward for loading. In some instances a combination trigger guard and lever was used. One of these pistols, sold by Parke-Bernet Galleries, New York, April 21, 1944, is pictured in Figure 80. In Gerhard Bock's *Moderne Faustfeuerwaffen* is pictured another of these lever-operated pistols like the one illustrated, except in engraving. Bock notes that this has the spring-type extractor.

The Sharps pepperboxes in the following descriptions are of three distinct types. Two of these types are each in two distinct models. We have, therefore, a total of five models, not counting the already described and very rare lever-action Sharps. Each of these models is found with minor variations.

Separation into five models only is simply my own classification. Another student may, at his discretion, call each variation a separate model. He may then have from six to fifty or more models, depending upon where he wishes to stop in rating a minor difference as a distinct model.

In Chapter 4 a long array of Allens was pictured, showing small variations. An attempt to picture all variations of Sharps will not be undertaken. Variations that occur in one model usually occur in another model. It will suffice to give short descriptions of the variations and to refer the reader to pictures of other models in which the specific variation will be shown.

The first type was produced by Christian Sharps in two models, one being .22 caliber, Figure 81, and the other, larger, .30 caliber, Figure 82.

In comparing the two pictures it will be noticed at once that the grips are of different materials and that the ratio of barrel length to frame length is different. In examining these two models, the student interested in variations will notice particularly the shapes of the standing breeches and the junctions of frames with grips. Nearly all the .22's found will be like the one illustrated except they will have embossed hard rubber grips instead of the ivory shown. These .22's may be referred to as the standard .22's. Nearly every .30 found will resemble the one pictured, except that the standing breech will be like that on the .22—straight instead of fluted.

Closer examination of the two will reveal that the hammer spring on the .22 is the two-prong type of the patent model, while the hammer spring on the .30 is a later one-piece type. On opening these two guns it will be seen that the .30 has a safety interlocking device between the hammer and the barrel release button, which is not found on this first, and by far the most common, .22.

Of a thousand Sharps of all types selected at random, perhaps four hundred will be .22's and of these, three hundred will be like the one illustrated, but with floral rubber grips. (This estimate is only my guess, of course.) The remaining one hundred .22's will have standing breeches fluted, late-type hammer springs, the interlocking feature, and grips of wood whose junction with the frame is curved instead of straight. They will also have left-hand twist rifling, five-groove; instead of right-hand, six-groove. One variation has the barrel released by a button on the side rather than underneath the frame. The curved frame juncture and the side release button became standard on .32 caliber types and are shown in illustrations of those types.

These .22 and .30 models are generally marked in a circle around the hammer screw on opposing sides: C. SHARPS & CO. PHILADA. PA. and C. SHARPS PATENT 1859. A few .22's have the

Fig. 80 Sharps—four-shot.
(COURTESY PARKE-BERNET GALLERIES, INC., NEW YORK.)

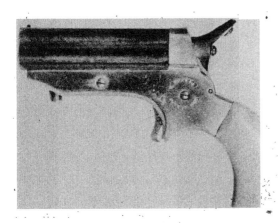

ig. 81 Sharps—5" overall—four-shot—.22 caliber—serials #52570.

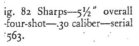

ig. 82 Sharps—5½" overall —four-shot—.30 caliber—serial 563.

markings on the barrel block just over the frame. Blued barrels and silvered brass frames are standard. A few .22's have been reported with iron frames. The .22's have 2½ inch and the .30's have 3 inch barrels, but on neither is the ratio of barrel length to frame length constant.

Of our hypothetical one thousand Sharps, four hundred were .22's. Another three hundred will be .30's, and of these perhaps two hundred will be simply larger versions of the .22 illustrated, but with checkered vulcanite grips. These last may be referred to as standard .30's. The remaining one hundred will have late-type hammer springs, safety interlock, left instead of right rifling, and nearly always rounded end grips. Probably none will have the side release button. Most will have the standing breech straight. A few will have a fluted standing breech, and of these last, one or two may be found like the one illustrated, with checkered grips and the union of frame and grips straight.

In our sampling we have accounted for seven hundred of the hypothetical thousand. The remainder will be about equally divided between the Sharps & Hankins and the Bull Dog types, both .32 caliber.

About 1863 Christian Sharps joined with William Hankins to form the firm of Sharps & Hankins.

The .32 Sharps & Hankin pepperboxes were probably made after the C. Sharps & Co. .22's and .30's of standard form, but before the variations of those models.

Two models of the Sharps & Hankins type are shown in Figures 83 and 84. Both guns are marked ADDRESS SHARPS & HANKINS, PHILADELPHIA, PENN. on top of the barrels. Both have 3½ inch barrels and use .32 long rim-fire ammunition. These cartridges fire a 90-grain bullet backed by thirteen grains of black powder. The .22's use .22 shorts holding four grains of powder with a 29-grain bullet; the .30's use .30 shorts, holding six grains of powder with a 55-grain bullet.

The Sharps & Hankins have been classified under two models only. They are divided arbitrarily by the location of one element, the revolving striker. Figure 83 shows a Sharps & Hankins

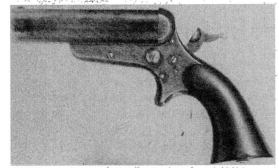

Fig. 83 Sharps & Hankins—6¼" overall—four-shot—.32 caliber—serial #3663.

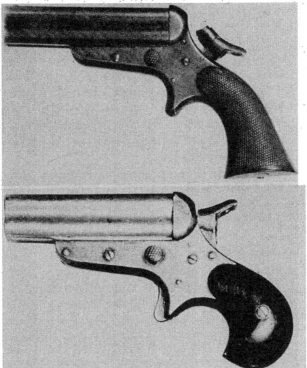

Fig. 84 (*center*) Sharps & Hankins—6¼" overall—four-shot—.32 caliber—serial #7075.

Fig. 85 (*bottom*) Sharps—5" overall—four-shot—.32 caliber—serial #13272.
(ROBERT ABELS COLLECTION.)

with this mechanism in the frame; Figure 84 shows another with the mechanism on the hammer.

The striker-in-the-frame models do not have the safety interlock. The one pictured has a large firing pin; others have small pins. Those with small firing pins have side plates around the hammer screws but lack the small screw under the standing breech. Some have barrel latches of different construction, and these will lack the small screw forward of the release button.

The striker-on-the-hammer models have the safety interlock. Some have a side plate around the hammer screw, and these lack the little screw under the standing breech. The striker-in-frame model was apparently made first but abandoned in favor of the striker-on-hammer model, and some of these striker-on-hammer guns are factory conversions from the first model.

Probably most Sharps & Hankins found will have a sliding-blade shell ejector placed vertically between the two right and the two left barrels. This seems to have been not altogether satisfactory and to have been discontinued about the same time the hammer-in-frame was.

Sharps & Hankins models show uniformity, in some details, that is quite lacking on the .22 and .30 Sharps. The frames are uniformly of iron, with rounded junctions between frames and grips. The barrels all have right-hand twist rifling with five grooves, and the release buttons are on the left of the frame. The release buttons vary slightly in construction and position. Grips may be of either plain or checkered vulcanite. The right of the frame is usually stamped "C. Sharps Patent Jan. 25, 1859."

The Sharps and Hankins partnership was dissolved about 1866. Sharps then put out the last type of the famous Four Shooters. This was the Model 2½, better known as the Bull Dog, which fired a .32 short using an 80-grain bullet with nine grains of black powder. It was supplied either blued or plated.

Figure 85 shows a typical specimen with 3 inch barrels and a pin for barrel stop in the side of the frame. The number of these guns made with 2½ inch barrels was greater than the number made with 3 inch barrels. Some specimens have a screw under the frame for the barrel stop. This model was the most nearly standardized. This piece is marked C. SHARPS PATENT, JAN. 25, 1859 on the right of the frame.

The Bull Dog is readily distinguished from the Sharps & Hankins by the rosewood bird's-head grips. In construction it follows the Sharps & Hankins pattern; but the barrels are shorter, the striker is always on the hammer, the frame rather than the barrel is recessed for the cartridge head, and there is never an ejector.

B. Kittredge & Co., Cincinnati, Ohio, advertised this model for $5.75 in *Turf, Field and Farm* on October 16, 1874, calling it the "Sharps 4 Shooter." On July 8, 1876, they advertised it for $5.50 in the *Army and Navy Journal*. Finally, on September 1, 1877, they brought the price down to $5.00 in their *Army and Navy Journal* advertisement. This last advertisement said: "After several hundred thousand .22 and .30 caliber (Sharps) pistols were made and sold . . . the inventor . . . produced a pistol . . . of much larger caliber. B. Kittredge & Co. bought the entire product of this pistol. We call it Sharps Triumph. It has all the advantages of cylinder pistols, and shoots with greater penetration . . . Over 15,000 sold by us. We have a few hundred left which we propose to sell at $5.00 for wood handle and plated frame . . . " It has been estimated the total production of Sharps was around 150,000 so the "several hundred thousand" .22 and .30 caliber Sharps mentioned may be questioned. It is a little surprising that Kittredge would use a rather large advertisement to sell a remainder of "a few hundred."

Chapter 7
OTHER AMERICAN CARTRIDGE PEPPERBOXES

NEXT TO THE Sharps, the Remington pepperboxes were most in demand. The Remington company put out only percussion-cap revolvers to compete with Smith & Wesson cartridge revolvers while the Smith & Wesson patent ran its course. They did not try, like some others, to compete by developing some hybrid revolver that would not infringe the patent, but they did offer cartridge pepperboxes.

These pepperboxes are of two types, both manufactured by Remington under patents taken out by William Elliot. The first type is the scarce revolving cylinder gun shown in Figure 86. This is commonly called the Remington Zig-Zag, and is marked MANUFACTURED BY REMINGTONS ILION, N.Y. and ELLIOT'S PATENTS AUG. 17, 1858, MAY 29, 1860 on opposite sides of the cylinder. The revolving of the cylinder is effected by movement of a stud in the zig-zag grooves to be seen in the illustration. The stud moves horizontally only, back and forth with the ring trigger. It goes ahead in one of the straight slots when the trigger is pressed forward. It slips into one of the slanting grooves when the trigger is pulled back, and so pulls the cylinder around. The backward pull on the trigger also cocks the hammer. When the stud passes from the slanting groove to another straight groove, the hammer is released. The .22 shorts are loaded through an opening back of the slanting breech.

Advertisements by T. W. Moore, New York, in 1862, in *Frank Leslie's Illustrated Newspaper* and in *Harper's Weekly* refer to this gun as "Elliot's Pocket Revolver." It is offered at $9.50 with blued frame and at $10.00 with plated frame. The gun is referred to as "the most compact, safe and powerful Pocket Revolver ever made." The further statement that it "will penetrate one inch of pine at 150 yards" is explicit. It is in contrast to another advertisement of the period which said a bullet from a Sharps would penetrate four inches of board, but did not name the wood or the range.

The second type is shown in Figures 87 and 88. This type is commonly called the Remington-Elliot. One model is a five-shot .22 with a round barrel group; the other is a four-shot .32 with a square group. The operating principle is the same on both, as is the marking—MANUFACTURED BY E. REMINGTON & SONS, ILION, N.Y. and ELLIOTS' PATENTS, MAY 29, 1860—OCT. 1,

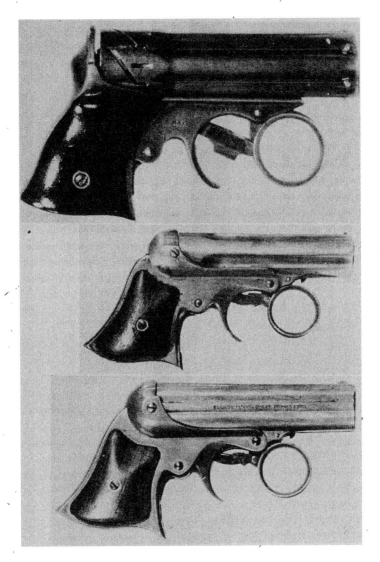

Fig. 86 (*top*) Remington—4¼″ overall—six-shot—.22 caliber—serial #827.
(GLODE M. REQUA COLLECTION.)

Fig. 87 (*center*) Remington—4¾″ overall—five-shot—.22 caliber—serial #9762.
(ROBERT ABELS COLLECTION.)

Fig. 88 (*bottom*) Remington—5″ overall—four-shot—.32 caliber—serial #22999.

1861 on opposite sides of the barrel group.

This type is quite different from the revolving cylinder type. The solid group of barrels drops for loading when a sliding catch ahead of the trigger is released. Pulling back the ring trigger, after pushing it forward between shots, revolves a striker in the frame by means of a ratchet and cocks the hammer by a lever. Final pressure releases the hammer.

Calling this second type "Elliot's New Repeaters," the Elliot Arms Co., New York, advertised (*Frank Leslie's Illustrated Newspaper*, Sept. 12, 1863) the .32 models as "the most safe, compact, durable, effective, sure and reliable Revolvers made." Elliot's justification for his use of the first of the array of adjectives was the incorporation of an unusual safety in his invention. The edges of the rim-fire cartridges used in all the cartridge pepperboxes contained the fulminate. It was important that they be guarded from accidental blows. Elliot's safety device held the striker away from the cartridge heads by a light spring until the hammer fell.

Pictured, Figure 89, is a Starr button-trigger, single-action cartridge pepperbox, patented by Eben Starr in 1864, Patent #42698. Mr. Starr took out an earlier patent, #14118, Jan. 15, 1856, which described and pictured a percussion-cap pepperbox, but he applied the features of this earlier patent to a revolver instead.

Mr. Starr's patent for the cartridge arm covered the revolving mechanism of the striker, and the barrel lock. Cocking the hammer rotates a sliding plunger which is driven by the hammer against a cartridge rim when the button-trigger is pressed. Thumb pressure on a bolt in the frame disengages a stud integral with the barrel group so that the barrels may be tipped down for loading. When the gun is closed, it is locked by the stud held in the bolt by spring pressure.

Another model, or variation, is shown in Figure 90. This has longer barrels, thicker grips, an external spring to hold the plunger mechanism in place, and a trigger with concave instead of convex nose. Another variation is shown in Figure 91. This has the external plunger spring and trigger with concave nose,

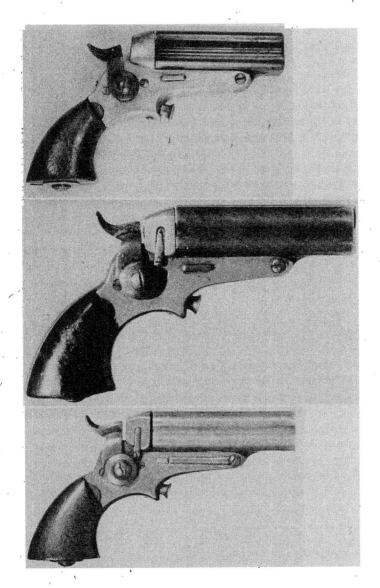

Fig. 89 (*top*) Starr—5½″ overall—four-shot—.32 caliber.
(DR. PRENTISS D. CHENEY COLLECTION.)

Fig. 90 (*center*) Starr—6″ overall—four-shot—.32 caliber—serial #68.
(GLODE M. REQUA COLLECTION.)

Fig. 91 (*bottom*) Starr—5¾″ overall—four-shot—.32 caliber—serial #22.
(FRANK R. HORNER COLLECTION.)

but it differs from the variation in Figure 90 in that it has an external barrel-release spring. It also differs from most Starr pepperboxes in having the marking STARR'S PAT'S MAY 10, 1864 on the left of the frame under the barrels. This marking is usually around the side plate on the left of the frame. Examination of the patent drawing does not suffice to show which of these models is the earliest. The drawing shows a convex trigger, an external bolt spring, and no external plunger spring.

None of the three Starrs pictured has a cylinder recessed for cartridge heads. However, some Starr pepperboxes, undoubtedly later models, have cylinders recessed.

An 1865 issue of *Frank Leslie's Illustrated Newspaper* carries an advertisement by Merrill Patent Firearms Co., Baltimore, Maryland, picturing a Starr four-shot. Some of the advantages claimed for this four-shot pistol are: "It carries a heavier Cartridge than any pistol of the same size," "Safest Pocket-Pistol Made," and "It has less parts than any other pistol."

A cartridge pepperbox put out by a maker who earlier produced percussion pepperboxes is the Bacon, pictured in Figure 92. This is a six-shot single-action with a cylinder revolved by cocking the hammer. The cylinder is removable for loading by unscrewing the center pin which also serves as an ejecting rod. It is marked on the cylinder BACON ARMS CO. NORWICH, CONN.

Mr. Jacob Rupertus, of Philadelphia, Pennsylvania, secured Patent #43606 on July 19, 1864, for a single-action cartridge pepperbox having several new features. One of these rather scarce pepperboxes is shown in Figure 93. By cutting away excess metal and by having the barrels converging to the front, rather than parallel, an eight-shot weapon of light weight was produced. There is a round breech-piece that may be turned through 90 degrees. This is notched at one point on the edge to permit the hammer to strike a cartridge rim. A loading gate is part of the breech-piece. This may be snapped open, for insertion or ejection of cartridges, when the breech-piece is turned through a quarter circle counter-clockwise. When thus turned, the breech-piece acts as a safety by barring the hammer from contact with

Fig. 92 Bacon—5½″ overall—six-shot—.22 caliber—serial #4.
(GLODE M. REQUA COLLECTION.)

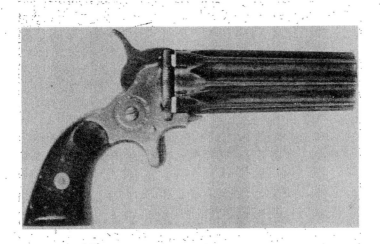

Fig. 93 Rupertus—5½″ overall—eight-shot—.22 caliber—serial #2688.

a cartridge. The ejecting rod, carried in an opening at the axis of the barrel group, has a springy split end which permits its being held in place by friction. Aiming is effected, according to the patent wording, by means of the longitudinal ribs on the barrels.

A pepperbox oddity is pictured in Figure 94. This knuckle-duster type is commonly called My Friend and is marked MY FRIEND PATD DEC. 26, 1865. The patent, #51752, was obtained by James Reid of Catskill, New York. The patent covered only the formation of the ring in the all-metal frame and a sliding safety designed to lock the gun with the hammer between cartridges. It is necessary to turn out the center pin and remove the cylinder to load it.

The gun illustrated in Figure 94 is a .22 without the safety. The guns are found mostly in .22 caliber, sometimes in .32, rarely in .41. The safety feature is usually omitted, except in .41 caliber.

Figure 95 is of a .22 with the safety. The safety consists of a sliding stop, movable in a groove in the under side of the frame. Its forward end is curved up to terminate in a round stud that may be pushed in a barrel, when at its lowest point, through an opening in the muzzle shield. This shield otherwise covers all the barrels except the one in line with the hammer. The safety will obviously lock a gun with an uneven number of barrels, with the hammer between the cartridge heads. Mr. Reid, according to his specification, contemplated only cylinders with an uneven number of barrels. The .22 is seven-shot, and the .41 is five-shot. The only .32's I have seen are five-shot also.

My Friend is shown in one of the figures in the patent drawing "as grasped in the hand as a means of defense, should the same be used in that manner instead of being fired." The illustration shows the little finger of the right hand passed through "the ring or bow," with the other fingers and the thumb gripping the cylinder. Mr. Reid in his specifications made other references to the use of the weapon as a knuckle-duster, but always "for protection" or "in hitting a blow for self-protection."

The July, 1923, issue of *Stock and Steel* reported that a

Fig. 94 (*top*) Reid's "My Friend"—4" overall—seven-shot—.22 caliber.

Fig. 95 (*center*) Reid's "My Friend"—4" overall—seven-shot—.22 caliber.
(HAROLD G. YOUNG COLLECTION.)

Fig. 96 (*bottom*) Continental Arms Co.—5¼" overall—five-shot—.22 caliber—serial #687.

keeper in Sing Sing suggested to Mr. Reid that a barrel added to My Friend would provide a better grip when striking a blow. The addition of a barrel makes the gun either a quasi-pepperbox or a revolver, as you wish. Two specimens with the barrel added will be shown in Chapter 10.

Less than a year after My Friend came on the market, the Ladies Companion appeared. It is pictured in figure 96. Although manufactured by Continental Arms Co. of Norwich, Connecticut, it was patented by Charles A. Converse and Samuel S. Hopkins, and the patent was assigned to the Bacon Manufacturing Company. The patent is #57622 of August 28, 1866. The marking on the cylinder is CONTINENTAL ARMS CO. NORWICH CT. PATENTED AUG. 28, 1866.

The Ladies Companion appears to have a round cylinder separated from a fluted cylinder. Actually, there is only a single cylinder, part fluted and part round. The invention "provided a better support for a pepper-box cylinder." The cylinder turns on a short center pin, integral with the frame, and is free to revolve in a ferrule which is screwed to the frame. The ferrule rests against a shoulder cut in the cylinder just forward of its round section. The cylinder is prevented from any movement other than turning on its axis. The recoil head is cut to permit loading of cartridges, but there is no ejector to clear the barrels for reloading. However, one remembers Richard Harding Davis' advice to a correspondent: "Carry a revolver and six cartridges. You will never need a gun more than once and if you don't get the other chap in six shots, he will get you."

A twentieth-century American cartridge pepperbox is the Shatuck Unique palm pistol, illustrated in Figures 97 and 98. This "squeezer pistol" was patented Dec. 4, 1906, Patent #837867, by Oscar F. Mossberg, Chicopee Falls, Massachusetts, and bears the marking UNIQUE C. S. SHATUCK ARMS CO. HATFIELD, MASS. It is held in the palm of the hand and fired by pressing a sliding part of the frame until a concealed hammer is tripped. The four short barrels are bored in a steel block which drops for loading when a catch is released. The striker is in the frame and is rotated by movement of the unconventional trigger. A

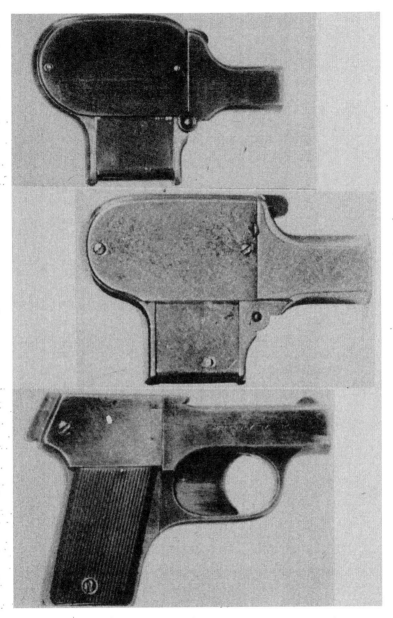

Fig. 97 (*top*) Shatuck Unique—3½" overall—four-shot—.22 caliber—serial #455.

Fig. 98 (*center*) Shatuck Unique—3¾" overall—four-shot—.32 caliber—serial #196. (A. D. MOWERY COLLECTION.)

Fig. 99 (*bottom*) Mossberg "Brownie"—4½" overall—four-shot—.22 caliber. (ROBERT ABELS COLLECTION.)

very narrow, flat, and thin steel bar is provided for ejection of shells. This slips into a small opening in the right side of the frame and is carried under the side plate with only the tip visible. Figure 97 is of the small .22 model. This has a blued iron frame and is fitted with a barrel catch placed vertically in the frame. Figure 98 shows a .32 which is considerably heavier and of slightly different shape. The .32 has an iron frame which is nickel-plated over an under-plate of copper, and on this the barrel catch is fitted horizontally.

The last of the pepperboxes is probably the Brownie, patented July 27, 1920. This not only was patented by Mr. Mossberg—as was the Shatuck Unique—it also was manufactured by him. The example shown in Figure 99 is marked O. P. MOSSBERG & SONS, NEW HAVEN, CONN. on the barrel and BROWNIE on the frame. Pressing a lever on the rear of the frame permits the hinged barrel block to be tipped down for loading of .22 rimfire cartridges in the four barrels. The rotating striker mechanism is similar to that in the Unique, but in the Brownie the trigger is conventional.

Chapter 8
EUROPEAN PERCUSSION PEPPERBOXES

PERHAPS THE American maker, with his knowledge of how accurately a Kentucky rifle would shoot, realized better than the European the shortcomings of pepperboxes. Probably the American maker was unwilling to put top quality in a piece he knew would be, at best, inaccurate, of low power, and of short range. The British maker, on the contrary, though he may have been fully aware of the ineffectiveness of a pepperbox, was still willing to put fine workmanship in it. Elaborately engraved frames and finely checkered grips do not make a gun a better weapon. Such adornment, commonly found on British pieces, was not applied to American pieces. The Forty Niner or the traveler in America bought his pepperbox for a time of need and was little concerned about its beauty.

The metal parts of American pepperboxes were usually uniformly blued. The cylinders and frames of British pepperboxes were usually case-hardened to give a mottled appearance with shadings of blue, green, brown, and yellow. The engraved decoration of frames, barrels, guards, hammers—in fact, all metal parts—was often of high artistry, intricate and with fine detail. A fair number of British pepperboxes were made with frames of German silver.

Grips on American pepperboxes were, rather rarely, made of ivory, pearl, or tortoise shell, plain or sculptured. Most grips were of walnut, though rosewood, mahogany, and other beautiful woods were sometimes used. American wood grips were smooth. British grips were nearly all of walnut, usually checkered, the checkering being frequently of extreme fineness, with up to fifty parallel lines to the inch. On American pepperboxes where nipples are at right angles to the cylinder axis, they are usually in recesses countersunk in the cylinder. On British pepperboxes, recesses are very rarely provided.

Probably the reason we find few British pepperboxes patented is because the British makers considered as known to be in existence, and therefore not patentable, the elements of construction and design, such as the double-action lock and the mechanically rotated cylinder.

An early and very unusual British pepperbox is the Budding. Two models of this are shown in Figures 100 and 101. The first model, the one without a trigger guard, bears no proof marks. The other, and later, model has Birmingham proof marks. Mr. J. N. George says the Buddings are the least common of the British pepperboxes and estimates the date of their manufacture as 1825-30. Nothing is known definitely as to where, when, or by whom the Budding was made. Surprisingly, the label in a cased Budding does not give the maker's address. Major Brown, the British collector, inclined to the belief that Budding was an Irish maker.

The maker of this first model Budding discarded accepted methods of construction and produced a well-made weapon with a minimum of parts. The striker is a plunger integral with the trigger and is cocked by pulling back the trigger in a slot until it reaches a notch in the right of the slot into which it is turned. Pulling back this plunger compresses a heavy coiled spring. To fire, the trigger is pressed to the left out of the notch, and the spring drives the plunger against a percussion cap on the bottom barrel. The one-piece, hand-revolved bronze cylinder has nipples entirely enclosed (where they can't fall off) and set in line with the barrels. To cap the nipples it is necessary to remove the cylinder by withdrawing a center pin which is screwed into the frame. The center pin serves as a rammer. A catch in the top of the frame engages a notch in the cylinder to effect alignment of hammer and barrel.

In addition to the cocking trigger, the later model illustrated

Fig. 100 Budding—8" overall—five-shot—.31 caliber.

Fig. 101 Budding—8" overall—five-shot—.31 caliber.

has a conventional trigger for firing. To fire this piece, the cocking trigger is brought straight back to full cock, and the firing trigger is then pulled. The center pin is now integral with the frame. Upon removal of a holding screw, the cylinder may be slipped off the center pin. In both models the use of a screwdriver is necessary in order to recap the nipples. Both guns are marked only BUDDING MAKER.

Another unusual pepperbox of the same period is shown in Figure 102. This is known as the Rigby type and is found mostly in a much smaller size with three barrels. This gun, marked W. JACKSON and LONDON has four unscrew barrels, a folding trigger, and a revolving striker on the hammer. The striker hits the nipples in turn and is part of a revolving plate that is turned manually clockwise through 90 degrees between shots. A smaller pepperbox of the same type made by J. Collins, London, is the three-shot pictured in Figure 103. The type has been reported in six-shot. Any specimen is rare.

The Buddings and the "Rigbys" have peculiar firing mechanisms. When we come to British percussion pepperboxes of the ordinary forms, we find nearly all to be double-action and we seldom come across a single-action. A rare, early single-action British pepperbox is illustrated in Figure 104. This has a trigger that snaps open when the hammer is cocked and barrels that unscrew for loading. The use of an unscrew barrel and a folding trigger was common on single-shot pistols, but pepperboxes were rarely so made. This one can hardly be considered a primitive; it is finely made and engraved, with a trap in the butt for caps. It is marked R. EYKYN.

Two other unusual single-action, hand-revolved pepperboxes —one English, Figure 105, and one Irish, Figure 106—have the peculiar slanting nipples that are associated with Irish pieces of the early percussion-cap period. A nipple is commonly set so its vent is in a plane that bisects the barrel bore, and nearly always it is either in line with or perpendicular to the axis. Such a vent cuts the powder chamber in a circular opening. The nipples in these two pepperboxes are set in a slant with the vents pointing away from the barrel axes. The vents therefore cut the

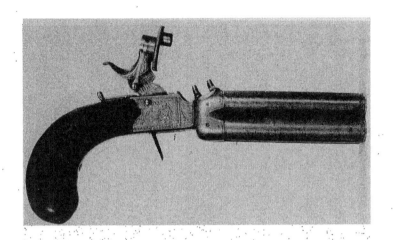

Fig. 102 Jackson—10" overall—four-shot—.42 caliber.

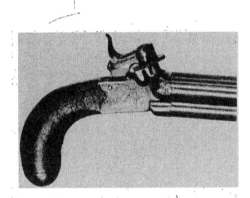

Fig. 103 J. Collins—4" overall—three-shot—.36 caliber.

chambers in elliptical openings, with the result that the openings are increased in area. The idea must have been to give a "fatter spark." The side hammers are bent so they will strike the canted nipples squarely. To use the target-shooter's terminology, the barrel in firing position is about "at two o'clock." Each gun has a solid cylinder and carries a loading rod screwed in its center. The English piece, marked SMITH LONDON, has sights and a cylinder that revolves clockwise, while the Irish gun, marked CUTLER, has no sights and a counter-clockwise cylinder. Both are of simple and very sturdy construction.

Another unusual and probably early pepperbox is pictured in Figure 107. This is quite unmarked and may be American or Continental, but it is included with British pieces because Joseph Manton is known to have made eighteen-shot pepperboxes of similar appearance. However, eighteen-shot pepperboxes marked "Joseph Manton" had only six nipples and fired three shots at a time. This gun has eighteen nipples connecting with eighteen barrels which are in two concentric circles in the cylinder, twelve barrels being in the outer row. The lock is double-action and fires one inner barrel after firing two outer barrels.

Figure 108 is of an unmarked—except for Birmingham proof marks—pill-lock pepperbox that almost surely was made by J. R. Cooper. Another pepperbox, not illustrated, marked J. R. COOPER, almost identical to this and including a hammer that is not reinforced, is known to exist in percussion-cap ignition. The folding trigger on this pepperbox is not common, but the method of ignition is what makes this rare. The nose of the hammer crushed pellets of fulminate to detonate the powder. Pill-lock ignition was often used on arms of other types, but not on pepperboxes.

Comparison of the illustration of the unmarked pill-lock pepperbox with Figure 109 shows some similarity. This latter pepperbox is marked J. R. COOPER and has percussion-cap ignition. On this the hammer nose is reinforced. It is to be noted that both these Cooper pepperboxes are double-action with bar hammer and mechanically turned cylinders.

Most Cooper pepperboxes are of the ring-trigger, underham-

Fig. 104 (*top*) R. Eykyn—9" overall—four-shot—.34 caliber.
(W. KEITH NEAL COLLECTION.)

Fig. 105 (*center*) Smith—7½" overall—seven-shot—.32 caliber.
(FRANK R. HORNER COLLECTION.)

Fig. 106 (*bottom*) Cutler—8" overall—five-shot—.40 caliber.
(FRANK R. HORNER COLLECTION.)

mer variety, with the same mechanism as that of the Mariette System pepperboxes that were much in favor on the Continent. Mr. James Rock Cooper obtained a number of English patents from 1840 on. Some of the pepperbox patent drawings show a ring trigger, but none shows the other features of the standardized action. This action was used in Blunt & Syms pepperboxes and has been explained in Chapter 5.

Customarily, English pepperboxes with this Mariette action are marked "J. R. Cooper's Patent," but the one shown in Figure 110 is marked just J. R. COOPER. This is six-shot, as most were made, but it is a little unusual in the shape of the grips.

Another pepperbox, Figure 111, with only five barrels drilled in the cylinder, is marked J. R. COOPER'S PATENT—nothing else, save the usual Birmingham barrel proofs.

Another pepperbox, four-shot, made under the Cooper Patent by C. Maybury, is illustrated—Figure 112. This is a finely finished piece that is unusual in that the grips extend to the recoil shield. There is a sliding safety over the backstrap which when pressed forward moves between the nipples and prevents turning of the cylinder. The ring trigger is formed like a coiled snake. Marking on the backstrap is C. MAYBURY J. R. COOPER'S PATENT.

One of Mr. Cooper's numerous firearms patents was for a spring device to be connected with the hammer of a top-hammer pepperbox. This was designed to prevent the often ill-fitting caps from falling off the nipples. I do not know whether this invention was ever marketed.

A finely made and unusual pepperbox by J. Beattie, London, is that shown in Figure 113. The frame is of German silver and is stamped IMPROVED REVOLVING PISTOL. This designation is commonly given to the transition revolvers which were often nothing but conversions of pepperboxes (to be mentioned briefly in Chapter 10). The trigger folds forward and flies out when the button on the right of the frame is pressed. The cap box in the butt is a refinement often found, but the ring (original) on the tang is odd.

A late but very seldom found type of percussion-cap pepperbox is shown in Figure 114. It bears Birmingham proof

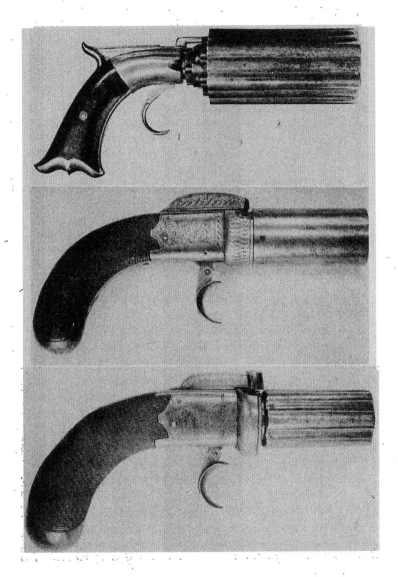

Fig. 107 Unmarked—9½" overall—eighteen-shot—.34 caliber.
(FRANK R. HORNER COLLECTION.)

Fig. 108 Unmarked—7¼" overall—six-shot—.31 caliber.
(DR. PRENTISS D. CHENEY COLLECTION.)

Fig. 109 J. R. Cooper—7½" overall—six-shot—.31 caliber.
(DR. PRENTISS D. CHENEY COLLECTION.)

marks but no maker's name. If it had a short cylinder and a barrel, it would much resemble a Deane-Adams revolver. It has these characteristics of a Deane-Adams revolver: (1) a double-action revolving mechanism of the revolver type, (2) a bolt locking the cylinder at the moment of firing by pressure against a partition, (3) horizontal nipples with partitions between them, (4) a flash-diverting shield over the cylinder, and (5) a spring safety that may be pressed in to hold a slightly raised hammer free from contact with a cap. This safety, standard on Adams and Tranter revolvers but unique on a percussion-cap pepperbox, is automatically released by pulling the trigger.

The great majority of English pepperboxes are of the double-action top-snap variety, having the standardized and unpatented action whereby the trigger pull synchronizes the turning of the ratcheted cylinder with the lifting and releasing of the hammer. Sometimes the trigger action also operates a bolt which holds the cylinder at the instant of firing. (If a pepperbox has this bolt, the cylinder is not free to turn when the trigger is fully back. Without the bolt lock, the cylinder will turn freely when the trigger is back.) When the pull on the trigger is released, the trigger is moved to its original position by a spring. The triggers are uniformly large and much curved; the hammers, sometimes called "bell-crank hammers" because they are essentially bell-crank levers, are uniformly of the bar type. All have solid cylinders which are held in position by a screw at the end of the center pin.

Just in case the names of various makers of these unpatented British pepperboxes should be desired, here is a partial list: Henry Allport, Baker, Beattie, W. A. Beckwith, Bentley, Berry, John Blanch, Bond, Thomas Boss, Collins, J. R. Cooper, Dooley, C. & H. Egg, Harkom, James Harper, T. & W. Harrison, Charles Jones, Charles Lancaster, Joseph Lang, Lacy & Co., Samuel Nock, Charles Osborne, Parker-Field & Sons, Owen Powell, Redfern & Bourne, Westley Richards, Richardson & Co., Rigby, Riviere, Smith, Southall, Henry Tatham, Wilkinson.

The following descriptions of top-snap pepperboxes will detail briefly minor differences commonly found among them.

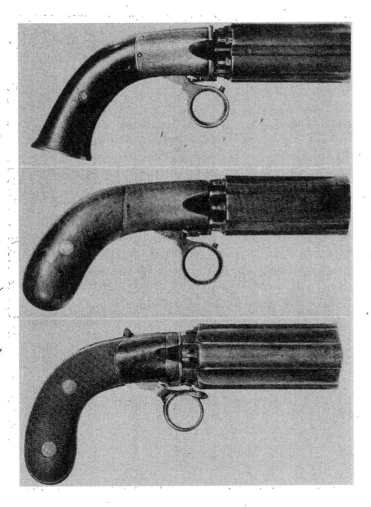

Fig. 110 (*top*) J. R. Cooper—8″ overall—six-shot—.38 caliber.
Fig. 111 (*center*) J. R. Cooper—7½″ overall—five-shot—.38 caliber.
Fig. 112 (*bottom*) C. Maybury—7¾″ overall—four-shot—.48 caliber.

110   EUROPEAN PERCUSSION PEPPERBOXES

Figure 115 shows a SMITH LONDON pepperbox with two-piece grips and a flat metal butt made as part of the frame. The mainspring is straight and heavy.

Figure 116 pictures an unmarked (except for Birmingham proof marks) pepperbox with a one-piece grip and a short V-spring. Though unmarked, it has engraving of better quality and more finely checkered grips than the previous example shows. Like most of the pepperboxes with one-piece grips, this has a silver monogram plate set in the top of the grip.

An early COLLINS REGENT STREET LONDON pepperbox is shown Figure 117. This was made without a nipple shield. On later pieces the nipple shield became standard construction.

Figure 118 is of a J. BLANCH LONDON pepperbox which has a steel butt cap, with a trap for caps, fitted on the one-piece grips. This has the form of safety most commonly used, a bar which slides under the partly raised hammer and keeps the hammer from contact with a cap. Pulling the trigger automatically takes the safety off by the hammer's backward pressure.

The finely made piece by BAKER LONDON, Figure 119, has a similar form of safety, but this safety can be released only by drawing it back intentionally.

The LACY & CO. LONDON pepperbox, Figure 120, has a still different safety. When pushed forward, it completely locks the action with the hammer firmly down until purposely drawn back. However, the hammer may be set down between caps. There is a projection on the under part of the nose of the hammer that fits in a shallow slot in the cylinder between and back of the nipples. Nearly all British pepperboxes have this fitting on the hammer.

American and Continental makers of percussion-cap pepperboxes apparently gave little thought to supplying lock safeties. British makers seem to have given the subject much study, but it is clear there was no agreement as to the most desirable form.

It is not surprising that we find the safety designed to meet the wishes of all on the very superior pepperbox, Figure 121, by WESTLEY RICHARDS. This is a beautifully engraved example of the ingenuity and craftsmanship of a top British maker. The

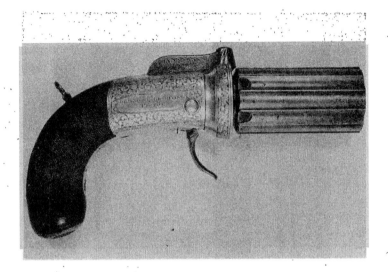

Fig. 113 J. Beattie—8″ overall—six-shot—.40 caliber.
(DR. PRENTISS D. CHENEY COLLECTION.)

Fig. 114 Unmarked—6¾″ overall—six-shot—.35 caliber.
(FRANK R. HORNER COLLECTION.)

sliding safety may be used to lock the gun entirely, or it may be set simply to raise the hammer so it will not touch a cap. When set only to keep the hammer from accidental contact with a cap, this safety is automatically released when the trigger is pulled:

The pepperbox by JOSEPH LANG HAYMARKET, LONDON, Figure 122, differs from the run-of-the-mill English top-snap pepperboxes in several respects. It is large and heavy and of exceptionally large caliber. The frame is of German silver and the grips, instead of being checkered, are left smooth, showing the beauty of the wood. The cylinder has partitions between the nipples. This simple protective feature is rarely found on English top-hammer pepperboxes. Any pepperbox with that feature was probably made after the 1851 expiration in England of the Colt patent.

A "detective special" pepperbox is shown in Figure 123. This one, by OWEN POWELL SHEFFIELD, is unusual by reason of its cylinder being only two inches long.

Nearly all the top-snap English pepperboxes were six-shot. A five-shot, made by T. BOSS LONDON, is shown in figure 124; and a four-shot, by JOHN COLLINS LONDON, in Figure 125.

Underneath views of two pepperboxes are shown in Figure 126. These are conventional in construction except that one, made by LACY & CO. LONDON, is fitted with a belt hook, and that the other, unmarked, has a lion's-head butt and fancy trigger guard.

The care taken in drawings in English patents is noticeable in the reproduction in Figure 127 of an 1845 patent drawing. The drawing may be of an actual pepperbox, or it may be only an artist's design. Pepperboxes having as fine workmanship were often produced, but pepperboxes embodying the curious patented features of the one pictured were manufactured in very small numbers, if at all. The unique feature of the invention by Charles James Smith of Birmingham was a priming magazine fitted in the frame of a concealed-hammer pepperbox. The magazine was a thin circular container that functioned like one of the old percussion-cap "cappers," except that it fed pellets of

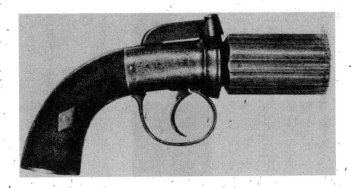

Fig. 115 Smith—7¼″ overall—six-shot—.36 caliber.

Fig. 116 Unmarked—7¾″ overall—six-shot—.37 caliber.

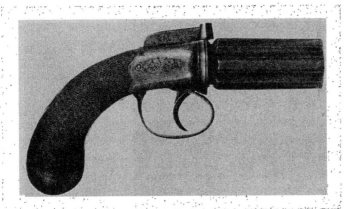

fulminate instead of caps. When a touch hole came in line with the hammer, it received a single pellet. The mechanism was complicated and probably not safe; so it is unlikely that a collector will come across a specimen.

Figures 128 and 129 show two English pepperboxes fitted with daggers. The one with the dagger screwed in the butt has a frame made entirely of metal. The other has an unusually long dagger, extending from the muzzle.

To remove the cylinder from the usual British pepperbox, it is necessary to turn out a screw in the end of the center pin. The pepperbox made by William & John Rigby, Dublin, Figure 130, is constructed so that the cylinder may be slipped off the center pin without use of a tool. This finely made and seldom found Irish piece has buttons on either side of the frame which, when depressed, release the cylinder.

An uncommon form of double-action English pepperbox with a mechanically rotated cylinder is shown in Figure 131. This has a superficial resemblance to the Cooper, but it is constructed quite differently. The hammer is entirely enclosed and strikes the top, rather than the lowest, nipple. The nipples are horizontal and separated by partitions. This type is known as the IMPROVED CENTRAL FIRE, and this piece is so marked. The frame, of German silver, also bears the maker's name THOS. CONWAY MANCHESTER. On the left of the frame is a button which may be pushed in when the trigger is drawn back slightly, to act as a safety.

An unusual unmarked pepperbox lacking even proof marks, which appears to be British, is shown in Figure 132. (English barrels may be expected to show either London or Birmingham proofs, but Irish and Scottish barrels often do not have proof marks.) Like the Conway pepperbox mentioned in the preceding paragraph, this is double-action with an internal top hammer. Unlike the Conway, it has Damascus barrels which unscrew. The breech to which the barrels are screwed is not readily removable from the frame, as it would normally be on a Continental pepperbox with screw barrels. The bore, .58 caliber, is unusually large.

We cross from England to the European Continent and find

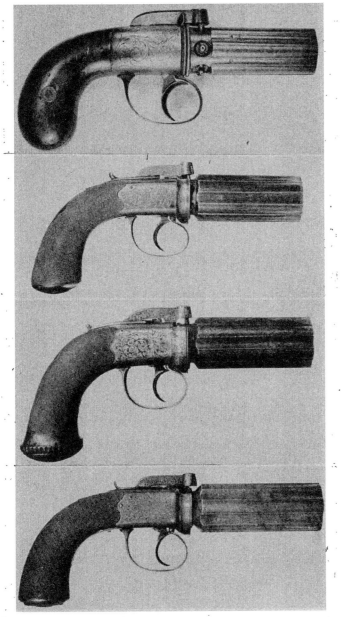

Fig. 117 (*top*) Collins—7″ overall—six-shot—.28 caliber.
(DR. PRENTISS D. CHENEY COLLECTION.)

Fig. 118 (*second from top*) J. Blanch—8½″ overall—six-shot—.41 caliber.

Fig. 119 (*second from bottom*) Baker—8½″ overall—six-shot—.38 caliber.

Fig. 120 (*bottom*) Lacy & Co.—9½″ overall—six-shot—.42 caliber.

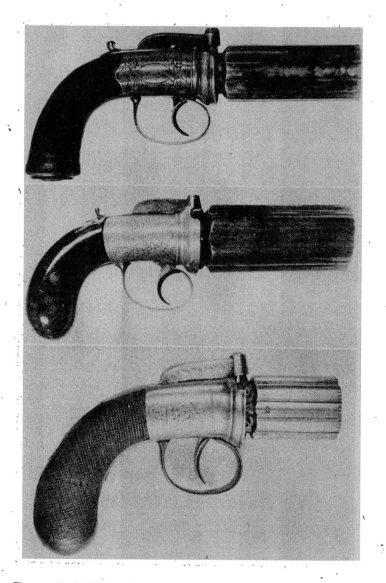

Fig. 121 (*top*) Westley Richards—9" overall—six-shot—.40 caliber.
Fig. 122 (*center*) Joseph Lang—9¾" overall—six-shot—.52 caliber.
Fig. 123 (*bottom*) Owen Powell—5½" overall—six-shot—.30 caliber.
(DR. PRENTISS D. CHENEY COLLECTION.)

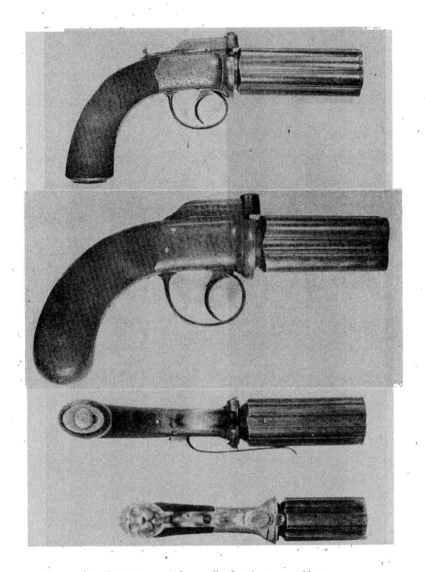

Fig. 124 (*top*) T. Boss—9½″ overall—five-shot—.50 caliber.
(DR. PRENTISS D. CHENEY COLLECTION.)

Fig. 125 (*center*) John Collins—8¼″ overall—four-shot—.48 caliber.
(DR. PRENTISS D. CHENEY COLLECTION.)

Fig. 126 (*bottom*) Underneath views of two British pepperboxes.

that pepperboxes of the Mariette system predominated. These were forerunners of the similar Cooper's Patent pepperboxes of England and of the Blunt & Syms pepperboxes of the United States. Their operation was explained in the description of the Blunt & Syms pepperboxes in Chapter 5.

In the United States a patent is given only to the first inventor, and it is fully disclosed. In France, Belgium, and other countries on the Continent, the patent was given to the first to register the application, at least until recent years. He was not required to be the first inventor, nor was he required to disclose the invention fully. Some European patents are very ambiguous. The patent ran from the date of its registration, and there was no examination to determine whether the invention had novelty. In the United States, the patents ran from the date of issuance and were not granted after 1836 unless "novelty" was shown.

Just when, where, and by whom the double-action, underhammer Mariette type was invented, I do not know. Some say 1837, but the date may be earlier. The European Mariettes were set apart, in form, from the American and English pepperboxes by their separate barrels which are bored through and screwed into the standing breech. The standing breech, containing the powder chambers and the nipples, may be detached from the frame, after the barrels are turned off, by unscrewing a bolt. The double-action mechanism, which employs a ring trigger, rotates the standing breech and the barrels with it. This mechanism would work with a great number of barrels. Mariettes with as many as twenty-four barrels were made. Some cartridge Mariette Brevete pepperboxes do not have separate barrels, nor do they have Damascus barrels. However, Damascus barrels are a distinctive feature of percussion-cap Mariettes.

To produce the very fine and ornamental figure on the best Damascus barrels, thirty-two layers, alternately of iron and steel, were rolled, cut into rods 3/16 of an inch square in cross section, and twisted until there were up to eighteen turns to the inch. As many as six of these rods, some with right twist and some with left twist, were joined at welding heat and rolled into a ribbon. The ribbon was then spirally coiled and welded on

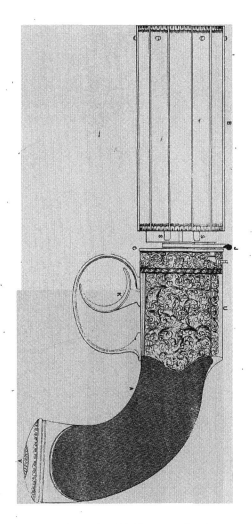

Fig. 127 Reproduction of an 1845 British patent drawing.
(COURTESY RAY RILING.)

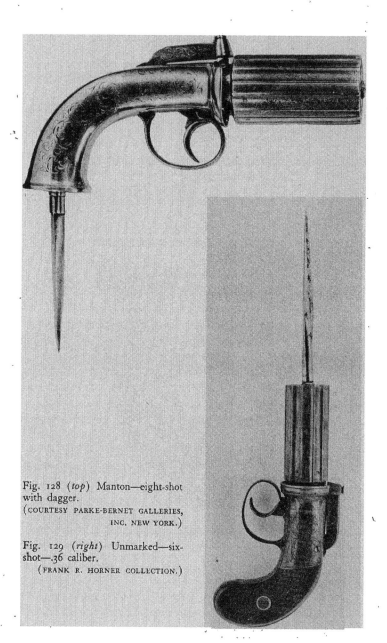

Fig. 128 (*top*) Manton—eight-shot with dagger.
(COURTESY PARKE-BERNET GALLERIES, INC. NEW YORK.)

Fig. 129 (*right*) Unmarked—six-shot—.36 caliber.
(FRANK R. HORNER COLLECTION.)

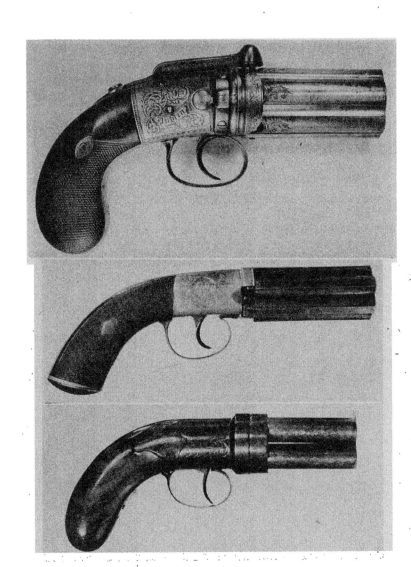

Fig. 130 (*top*) Rigby—7″ overall—six-shot—.38 caliber.
(DR. PRENTISS D. CHENEY COLLECTION.)

Fig. 131 (*center*) Conway—9″ overall—six-shot—.41 caliber.

Fig. 132 (*bottom*) Unmarked—8¾″ overall—four-shot—.58 caliber.
(FRANK R. HORNER COLLECTION.)

a mandrel. The barrel was pickled in acid and burnished. The effect was a barrel surface covered with minute curving lines, alternately bright from the steel and dark from the iron. The silver tone of these barrels was combined with both bright steel and gray steel in the frame, and with ebony or vulcanite grips, to produce a silvery black and white effect.

The pepperboxes of Continental Europe range from very ordinary to very elaborate and beautifully made pieces. Some were cheaply made to be sold at a low price. Others especially made by the top makers of France and Belgium were works of art that sold at high prices. These sometimes have gold inlaid frames and barrels and may even have gold inlaid screw heads. Belgium has produced many of the poorest and cheapest guns ever made, but Belgium has also produced some of the finest rifles, pistols, and pepperboxes ever manufactured.

Mariettes differ in size, in shape, and in the number of barrels, but not in general construction. Some are bulky; others are slim and graceful. Figures 133 to 137, inclusive, picture a number of Mariettes. The first is notable for its smallness. However, it is only small, not miniature. The next is a slim piece of small caliber. The rest represent the usual run of Mariettes, having from four to eight barrels.

Figure 138 shows a more striking Mariette, remarkable in that it has eighteen barrels, twelve in the periphery and six in an inner concentric circle. The eighteen nipples are arranged in a circle, the vents leading straight to the outer barrels and curving to the inner barrels. Successive trigger pulls will fire two outer barrels and then one inner barrel.

Figure 139 shows a Mariette that is exceptional because of its quality. The beautiful Damascus barrels are numbered and the numbers are in gold inlaid designs. All the metal furniture is finely engraved and, even to the screw heads, is gold inlaid. The grips are of ivory, fluted and carved. In gold on the backstrap is M. J. CHAUMONT, FABRICANT A LIEGE. No joint is seen in the butt, though the butt plate turns to disclose a cap box.

Perhaps it should be mentioned that the four grooves in the muzzle of a Mariette pepperbox barrel are sometimes mistaken

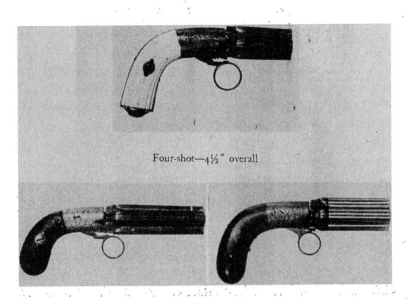

Four-shot—4½" overall

Four-shot—7" overall   Four-shot—7" overall

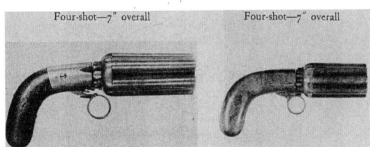

Five-shot—7½" overall   Eight-shot—7" overall

Figs. 133 to 137, inclusive—Group of Mariettes.
(EIGHT-SHOT FROM FRANK R. HORNER COLLECTION.)

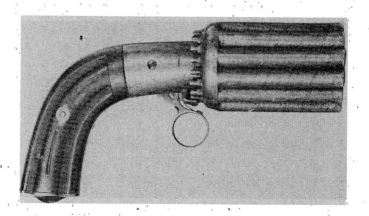

Fig. 138—Mariette—8″ overall—eighteen-shot—.31 caliber.
(FRANK R. HORNER COLLECTION.)

Fig. 139 Chaumont—9½″ overall—four-shot.

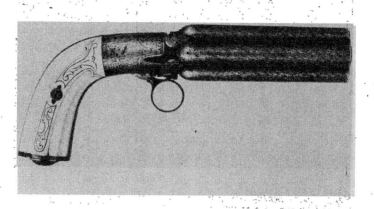

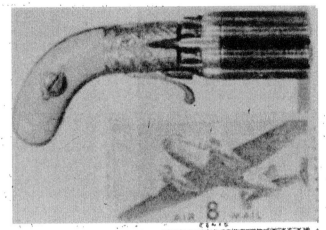

Fig. 140 Miniature—2⅛" overall—six-shot. Stamp more than 1½ times actual size. (DR. W. R. FUNDERBURG COLLECTION.)

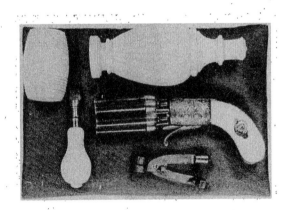

Fig. 141 Same miniature with accessories.

for rifling. These grooves are intended to hold the square end of the barrel wrench, which is inserted in the muzzle to unscrew the barrel.

A very remarkable piece of the finest workmanship is the miniature illustrated in Figures 140 and 141. It is shown along with a stamp (both gun and stamp appearing more than one and one-half times actual size) and also out of its little ebony case with its accessories. A miniature firearm is always much too small to be held in normal fashion, even in a baby's hand. Miniatures of repeating firearms which function in all respects are rare collector items. This miniature pepperbox has six unscrew blued barrels, a finely engraved frame, and ivory grips. It is precision-made and can be fired like a Mariette. In addition to the ivory flask, ivory barrel for caps, and ivory-handled screwdriver fitted in the case, there is another accessory which is a combination mold, nipple wrench, and barrel wrench. The mold is for a bullet about the size of a #7 shot.

Figure 142 is of an early French single-action pepperbox, perhaps made by Perrin and Le Page but not marked with a maker's name. The cylinder consists of four barrels welded to a central rib and fastened to a manually revolved standing breech. The cylinder is locked by a bolt out of the standing breech. The operation of this bolt by means of a flexible trigger guard is a device rarely used. Flexing the fore part of the guard withdraws the bolt. Also unusual is the fact that the barrel on the right is the one that is fired, and by a side hammer.

Another early but not primitive French pepperbox is pictured in Figure 143. It has a side-hammer, single-action lock. The Damascus barrels are separate but not designed to be screwed out of the breech into which they are fastened. The barrel group is turned by manual twisting. Considerable pressure is required in twisting, because it is necessary to force a spring catch out of a notch in a circular plate fixed to the rear of the breech. This very simple method of constructing an efficient indexing catch was probably not patented, but it seems to have been overlooked by other makers.

The versatile Herman of Liege was an early manufacturer of

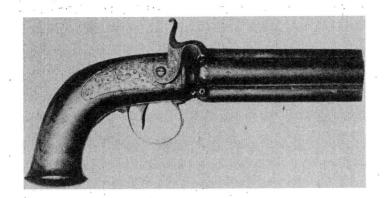

Fig. 142 Unmarked—9" overall—four-shot—.42 caliber.

Fig. 143 Unmarked—9" overall—four-shot—.33 caliber.

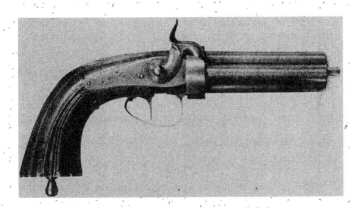

well-made pepperboxes. He seems to have been a man with many ideas, and not one to stop with drawing board design and an experimental model. He put his theories to trial in completed firearms.

A pepperbox by Herman is shown in Figure 144. It is double-action with a concealed top striker. The four barrels are screwed in, and turn with, the standing breech. The strange part is that the long center pin that passes through a collar in the frame is nothing but a camshaft. A cam rests in a slot in the standing breech and rotates it when trigger pull turns the camshaft by a lever. The figured barrels are plain twist, being made of a coiled ribbon of plain rods. The rods themselves were not twisted, as in true Damascus barrels.

Another Herman pepperbox with the same curious action is pictured in Figure 145. This is eight-shot and appears to be of later manufacture than the four-shot, with minor improvements. The chief improvement, the enclosing of the camshaft center pin in a sheath, makes the gun sturdier. The nipples, instead of being the type requiring a wrench with a square opening for removal, use the common form of two-pronged pin wrench. The trap in the butt has a turning friction cover instead of a snap cover such as the four-shot has. The barrels are now true Damascus, the trigger guard has a spur, and the grips are of ebony.

The Herman pepperbox shown in Figure 146 has a conventional double-action lock employing the ratchet method commonly used by other makers to turn the barrels. It has one difference that makes it decidedly unusual. Its eight barrels are arranged in the smallest group in which they can possibly be clustered. Each of the four outer barrels lies snug against two of four inner barrels. This Herman piece is marked J. J. H. BREVETE. Usually the full name is marked on Herman pepperboxes.

Still another Herman, Figure 147, has a peculiar lock but is otherwise of usual construction. The nipples are in line with the barrels. The top hammer is pivoted and turns up and back when being cocked, with the result that in falling it strikes a glancing blow on a cap.

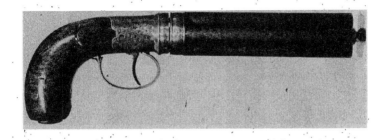

Fig. 144 Herman—10½" overall—four-shot—.36 caliber.

Fig. 145 Herman—9¼" overall—eight-shot—.32 caliber.
(FRANK R. HORNER COLLECTION.)

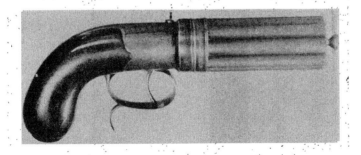

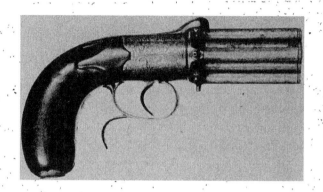

Fig. 146 Herman—8¼" overall—eight-shot—.30 caliber.
(FRANK R. HORNER COLLECTION.)

Fig. 147 Herman—8" overall—four-shot—.48 caliber.
(FRANK R. HORNER COLLECTION.)

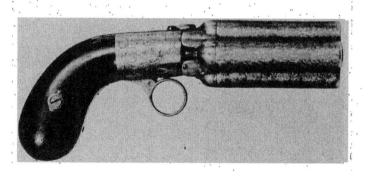

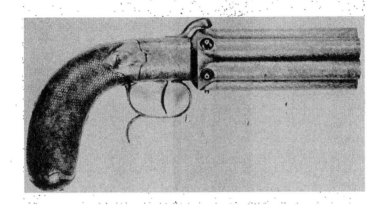

Fig. 148 T. Gannod—10″ overall—six-shot—.46 caliber.
(DR. PRENTISS D. CHENEY COLLECTION.)

Fig. 149 T. I. Hoist—7″ overall—six-shot—.38 caliber.

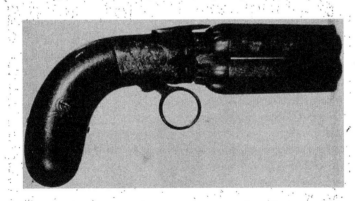

A most extraordinary firearm by Herman, not a pepperbox by definition, will be described in Chapter 10.

Figure 148 is of a well-made Swiss pepperbox with Damascus twist barrels that screw in the standing breech. The checkered grips of burl walnut and the double-action top-hammer mechanism are in the British mode, but the shape of the hammer is a departure from the normal. The frame is marked T. GANNOD A LAUSANNE, though the barrels have Belgian proof marks.

Figure 149 shows a ring-trigger pepperbox with separate barrels screwed to the revolving breech. It looks like a Mariette, but it is set apart from the Mariette by its action. The ring trigger operates a top hammer instead of an underhammer and rotates the cylinder by downward, rather than the usual upward, pressure on the cylinder tooth. The only marking except Belgian proof marks is T. I. HOIST stamped in small letters on the edge of the backstrap and fully hidden by the grips. As the peculiar cap over the hammer was Mr. Hoist's patent, the concealment of the name is unaccountable.

Another pepperbox which has some slight resemblance to a Mariette, but which is mechanically wholly unlike a Mariette, is pictured in Figure 150. This has needle-fire ignition. Long guns using needle-fire cartridges were made in great numbers. A considerable number of pistols, a few pepperboxes, and an even smaller number of revolvers were made for this ignition system. This particular piece is of fine workmanship and was made by Comblain in Brussels. The barrels are separate tubes, rifled and brazed together. The channeled ribs have a Damascus figure that contrasts with the closer twist of the barrels. The cylinder does not revolve. There are six firing pins struck in turn by a concealed revolving striker much like the Lancaster system. The means of revolving the striker is simple, effective, and very sturdy. Pulling the trigger draws back the shaft with its striker head and at the same time moves a cam follower against a cam on the shaft, thereby turning the shaft through one-sixth of a circle.

A needle-fire pepperbox of quite different construction is shown in Figure 151. It is marked RISSACK PATENT and has

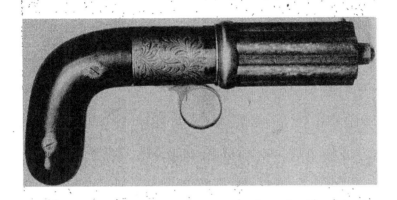

Fig. 150 Comblain—7½" overall—six-shot—.28 caliber.

Fig. 151 Rissack—7" overall—four-shot—.33 caliber.
(FRANK R. HORNER COLLECTION.)

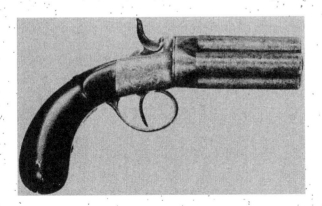

Belgian proof marks on the Damascus barrels. This one is single-action with a manually revolved cylinder. Turning the cylinder presses the needle back to receive the hammer blow. The barrels screw to a breech which is itself screwed to the frame, as in a Mariette.

Chapter 9
EUROPEAN CARTRIDGE PEPPERBOXES

NEARLY ALL British cartridge pepperboxes were made in just two types, the Lancaster and the Sharps. Except for the English Lancaster, all American and European pepperboxes are alike in not having the accuracy and striking power of first-rate revolvers. The Lancaster, a large weapon of pepperbox construction patented in 1881, was equal to the best revolver in accuracy and power, and superior in that it was less liable to become clogged or get out of order. These formidable guns are, even now, no more than obsolescent. Cartridges for some calibers are still being made. One of the early models is shown in Figure 152. This has four firing pins in the recoil head. A revolving and completely enclosed striker is turned by trigger pull to strike the pins in turn. The inclined groove rotating mechanism is much like that on the Robbins & Lawrence percussion pepperbox. A lever, pressed by the thumb, releases the solid barrel group so it may be turned down for loading. An extractor automatically empties the chambers when the gun is opened. The barrels are rifled on the Lancaster oval bore principle.

Later models have a single firing pin, integral with the striker. This improvement did away with the possibility of multiple explosions. With four pins there was the chance that recoil would drive the primer of the center-fire cartridge against a firing pin, other than the one under the striker, with sufficient force to fire the cartridge. The four-barrel Lancaster was made

Fig. 152 (*top*) Lancaster—10" overall—four-shot—.455 caliber.
(HAROLD G. YOUNG COLLECTION.)

Fig. 153 (*center*) Tipping & Lawden—4¾" overall—four-shot—.22 caliber.

Fig. 154 (*bottom*) Tipping & Lawden—7½" overall—four-shot—.33 caliber.
(DR. PRENTISS D. CHENEY COLLECTION.)

in .38, .45, .455, and .476 calibers. A .577 caliber was planned for the four-barrel, but it is doubtful that this large caliber was made for any but the Lancaster over-and-under two-barrel pistols. Variations in shapes of grips and in the forms of barrel-release levers—some operating from the side and some from the top—are numerous. One Lancaster model, rarely found, has the double-trigger mechanism found on some Tranter percussion revolvers. This system was condemned by men who had difficulty co-ordinating the movements of the forefinger and the middle finger but was well liked by others. An extension of the trigger protruded below the guard. Pulling this extension with the middle finger cocked the gun; pulling the regular trigger with the forefinger then fired it. By pulling with both fingers at the same time, the firing was the usual double-action. The incorporation of this double trigger—available at extra cost—may have been prompted by Lord Kitchener's only objection to the Lancaster—that it had too heavy trigger pull.

Here is some of the publicity given the Lancaster two-barrel pistol in the *Field Magazine* of June 10, 1882, and applying equally to the four-barrel pistol: "The advantages . . . over the old form of Revolver are, chiefly, increased accuracy and strength of shooting, owing to its dispensing with the escape (of gas) between the revolving chambers and the stationary barrel . . . There is in these Pistols no opening through which any gas can escape; so the weapon can be . . . (aimed) . . . with the left hand on the barrels, and there are no screws, hammers, or projections to catch on the clothing, reins, etc. Above all, there is no fear of a jam, or the weapon becoming unserviceable from overheating or other accident. The mechanism is covered up, and dirt, damp, wet, and ill-usage hardly effect these Pistols at all. (It has) a rebounding lock, rendering it safe to carry . . . "

Pocket-size pepperboxes made under the Sharps patent by Tipping & Lawden in Birmingham were moderately popular. The one illustrated in Figure 153 is a first type Sharps in construction, and also in appearance except that it has more ornamentation. This .22 has floral rubber grips, late-type hammer spring, and the interlocking feature. The junction of frame and grips is

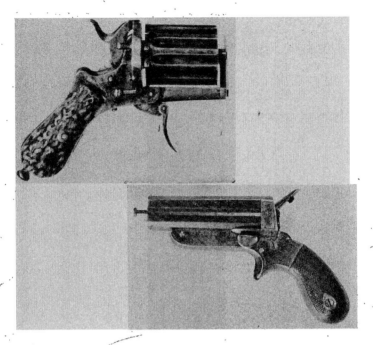

Fig. 155 (*top*) F. Hill & Son—5" overall—twelve-shot—.22 caliber.
(FRANK R. HORNER COLLECTION.)

Fig. 156 (*center*) Unmarked—5½" overall—four-shot—.32 caliber.
(DR. PRENTISS D. CHENEY COLLECTION.)

Fig. 157 (*bottom*) Sharps—Spanish copy—6" overall—four-shot—.32 caliber.
(ROBERT ABELS COLLECTION.)

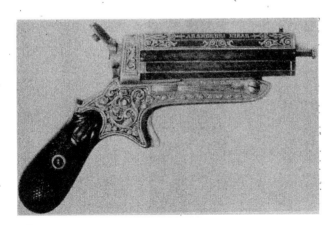

straight, and the standing breech is not fluted. The barrel release on this particular piece is a button under the frame, identical with the Sharps construction. Most Tipping & Lawden pieces will have a long spring, visible under the frame, as a barrel release. The marking on the left of the frame is TIPPING & LAWDEN SHARPS' PATENT.

One of the scarce Tipping & Lawden pieces of large size is shown in Figure 154. This uses a special rim-fire cartridge of about .33 caliber.

A pepperbox with British proof marks and bearing the name F. HILL & SON, SHEFFIELD on the backstrap is pictured in Figure 155. This type is usually called a fist pistol and on the Continent is only now becoming antiquated. It is double-action, with a folding trigger, a loading gate on the right, an ejector rod screwed in the butt, and two-piece bone grips. There is a spring safety of the Adams type on the left of the frame, designed to hold the hammer clear of cartridges. Most fist pistols are six-shot. This is twelve-shot.

There is much more diversity among European pepperboxes found on the Continent. Probably the best Continental pepperboxes are copies of the American Sharps. The copies show modifications, though not necessarily improvements. Sometimes the influence of Starr or Rupertus design may be surmised.

Figure 156 is of a pepperbox bearing neither maker's name nor proof marks. It appears to be of Central European, probably German, manufacture, and it possesses characteristics of both Sharps and Starr construction. A revolving firing pin turns as the hammer is cocked. The barrel group may be tipped down for loading after a button on the side is pressed. Cartridges are ejected when the gun is open by pressing in at the muzzle the long rod which runs through the center of the barrel block.

A Spanish gun with the same distinctive features is pictured in Figure 157. The frame is of chiseled steel and the barrels are gold inlaid. The marking on the barrel is FA DE JUAN ARAMBERRI EIBAR.

The Continental guns that bring to mind the Sharps pepperboxes are usually four-shot, but some are six-shot. Austrian

Fig. 158 Grunbaum—8" overall—six-shot—.31 caliber.
(FRANK R. HORNER COLLECTION.)

Fig. 159 Unmarked—8¼" overall—six-shot—.32 caliber.
(DR. PRENTISS D. CHENEY COLLECTION.)

Fig. 160 Sharps—Belgian—four-shot.
(MILWAUKEE PUBLIC MUSEUM COLLECTION
photograph courtesy *The Gun Collector*.)

pepperboxes marked GRUNBAUM'S PATENT WIEN, and SELBST-SPANNER were made in both four- and six-shot. One of the six-shot is illustrated in Figure 158. This has the revolving striker on the hammer but is unusual in having a double-action lock (a *selbstspanner* or "self-cocker") and a folding trigger. The barrels tip down to load, the release being a turntable lever which provides a positive lock.

Another unmarked Continental pepperbox that looks much like the Grunbaum is shown in Figure 159. This, however, is single-action with a stub trigger. It is closer than the Grunbaum to the Sharps construction. An ejector much like that found on Rupertus pepperboxes is screwed in the center of the barrel group.

The unusual double-action pepperbox, Figure 160, is marked C. SHARPS PATENT 1859 and bears Belgian proof marks. The barrels tip down for loading. The barrel release lever on the left of the frame locks the hammer when open.

Rim-fire cartridges, such as are used in Sharps pepperboxes, were brought to development in America. They represent improvements on the Flobert cartridges invented in France, which were at first merely bulleted breech caps. Pin-fire cartridges were well developed before Mr. Flobert's invention of the rim-fires. Guns using pin-fire cartridges and guns using rim-fire cartridges were displayed at the Great Exhibition of 1851.

An early form of cartridge pepperbox is the French pin-fire shown in Figure 161. This ring-trigger, double-action pepperbox has the center pin integral with the recoil plate. The barrels are welded together. The barrel group slides over the center pin and is held in position at the rear by a stud, and at the muzzle by a nut screwed on the center pin. The barrel group must be removed for loading. The recoil plate revolves by trigger action and turns the barrel group with it. The grips are one-piece, of carved ebony, and contain an ejecting rod screwed in the butt. The hammer has a fanciful dog's head as the striking nose, and under it the maker's name, BOISSY. The name can not be seen until the hammer is raised.

The Lefaucheux was probably the most popular of the early

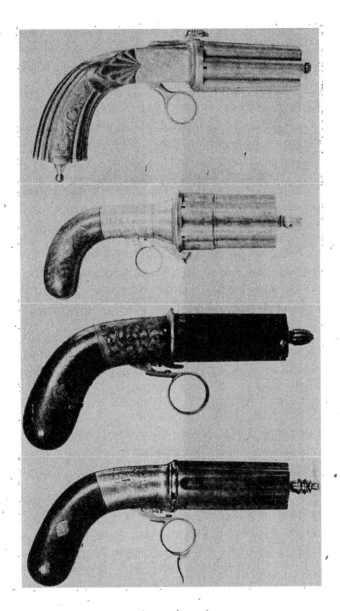

(*top to bottom*)

Fig. 161 Boissy—8″ overall—five-shot—.40 caliber.

Fig. 161A—Lefaucheux—8½″ overall—six-shot—.51 caliber.

Fig. 162 Mariette—6½″ overall—six-shot—.28 caliber.

Fig. 163 Flobert—7½″ overall—six-shot—.33 caliber.

pin-fire pepperboxes, but not many examples have survived. One is illustrated in Figure 161-A. The Lefaucheux is of distinctive form, with a long underhammer. The Boissy has a top hammer. Otherwise the mechanical features of the two pepperboxes have much in common. Both are ring-trigger and double-action, with the center pin integral with the recoil plate. Both have welded barrel groups that are secured and rotated in the same manner. The Lefaucheux illustrated has two-piece grips and no ejecting rod other than the center pin. It is marked INVon LEFAUCHEUX Bte ARQer ORDre DE MGR LE DUC DE NEMOURS.

The necessity of removing parts before reloading is never a commendable requirement. A part may be dropped and lost— or a part may be overtight and difficult to remove, as in the case of the box nuts used to secure the cylinders of both the Boissy and Lefaucheux pepperboxes. To permit the removal of a firmly tightened nut on a Boissy, a hole drilled through the nut permits insertion of a rod or nail to give leverage in turning. The box nut on the Lefaucheux has, in addition to the drilled hole, a kerf to allow the use of a screwdriver as an alternative.

Another pin-fire pepperbox is illustrated in Figure 162. This is marked MARIETTE BTE. and has the same mechanism as the percussion cap Mariettes. The nose of the underhammer is slightly changed so it will drive in the pins of the cartridges. The frame is of gray steel; the solid cylinder is blued. The center pin extends beyond the muzzle through an ornamented cap. When the cylinder is unscrewed for loading, the center pin, because of its extra length, will serve as an ejecting rod.

A pepperbox which at a glance would be mistaken for a pin-fire is shown in Figure 163. This is a rim-fire and is marked FLOBERT on the solid cylinder, MARIETTE BREVETE on the understrap. It is all nickel plated and has a supplemental folding trigger attached to the ring trigger. Otherwise it is like the last mentioned pin-fire, except that the hammer nose is designed to strike a cartridge head horizontally.

Another rim-fire pepperbox, Figure 164, is much like the last. This is marked only MARIETTE BTE. and has a blued cylinder and a bright frame. There is one marked difference—the

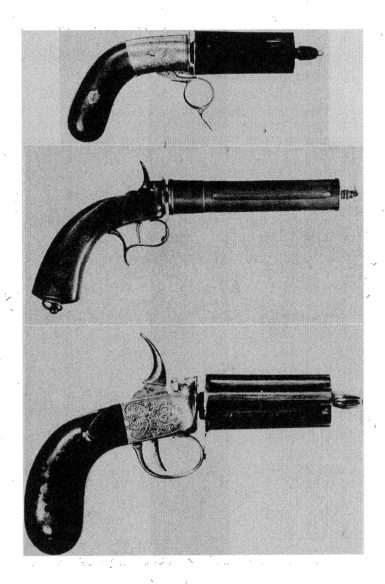

Fig. 164 (*top*) Mariette—7½" overall—six-shot—.32 caliber.

Fig. 165 (*center*) Mariette—12¾" overall—five-shot—.33 caliber.
(FRANK R. HORNER COLLECTION.)

Fig. 166 (*bottom*) Unmarked—7½" overall—five-shot—.32 caliber.
(HAROLD G. YOUNG COLLECTION.)

cylinder ratchet is part of a thin plate permanently attached to the rear of the cylinder.

Still another rim-fire pepperbox, also marked MARIETTE BREVETE, has none of the characteristics we expect in a Mariette. This, shown in Figure 165, is top-hammer with a conventional trigger, single-action with hand turned cylinder. The cylinder is turned by twisting and is indexed by means of a spring running from the trigger guard to notches cut in the cylinder breech. The striker is flat and wide, hitting completely across the cartridge head. This gun is unusually large.

A smaller pocket-size version is shown in Figure 166.

Pin-fire ignition was commonly used on the later fist pistols which attained considerable popularity and also on the pepperboxes which were occasionally made to be fitted in canes, bicycle handle bars, and purses.

Pin-fire fist pistols were made in quantity by various manufacturers. They were double-action with a forward folding trigger, usually of ordinary workmanship, produced to sell at a low price. Most of them do not bear a maker's name, but they do have Belgian proof marks. The firing mechanism is simple. The folding trigger is attached to a plate that is pivoted under the frame. When the trigger is drawn back, this plate turns, and in so doing it raises the hammer, pushes the cylinder around, and locks itself against the cylinder at the moment of firing. Several fist pistols are shown in Figures 167 through 170.

Figure 167 shows a Belgian piece of regular construction. This has a pivoted gate in the right of the recoil shield, permitting loading of cartridges without removing the cylinder. An ejecting rod is screwed in the butt. There is an incongruity in the fine polygroove rifling of barrels on a gun with no sights.

A smooth-bore Belgian piece with a loading gate on the left is pictured in Figure 168.

Figure 169 illustrates a fist pistol, marked with Belgian proofs and F. DECORTIS BREVETE. The cylinder must be removed to load with cartridges. The center pin is hinged to permit its outward tip to be secured by a bolt to the extension of the frame beneath the cylinder. Pressure on a spring releases the bolt so the

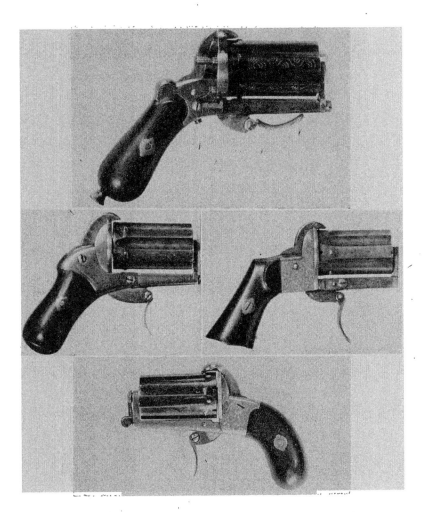

Fig. 167 (*top*) Unmarked—5" overall—six-shot—.29 caliber.

Fig. 168 (*center left*) Unmarked—4¼" overall—six-shot—.29 caliber.
(FRANK R. HORNER COLLECTION.)

Fig. 169 (*center right*) F. Decortis—4¾" overall—six-shot—.29 caliber.
(FRANK R. HORNER COLLECTION.)

Fig. 170 (*bottom*) Unmarked—4½" overall—six-shot—.32 caliber.
(DR. PRENTISS D. CHENEY COLLECTION.)

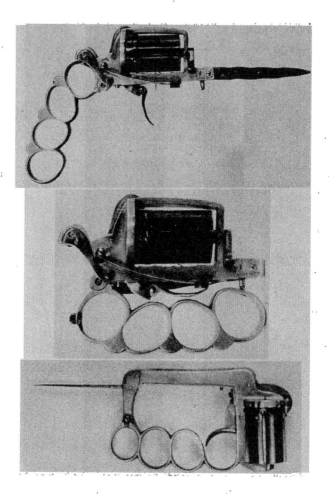

Fig. 171 (*top*) L. Dolne—8¾″ overall—six-shot—.29 caliber.

Fig. 172 (*center*) Another view of Fig. 171.

Fig. 173 (*bottom*) Unmarked—7″ overall—six-shot—.30 caliber.

(DR. PRENTISS D. CHENEY COLLECTION.)

Fig. 174 Devisme—5¾" overall—six-shot—.28 caliber.

Fig. 175 Unmarked—6½" overall—five-shot—.410 caliber.
(ANTHONY A. FIDD COLLECTION.)

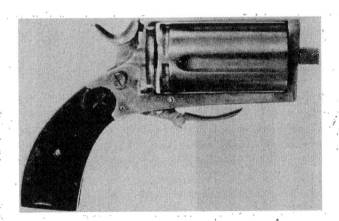

hinged center pin will straighten to allow the cylinder to slide from it.

Another piece with cylinder removable by withdrawal of a modified wing bolt is shown in Figure 170.

Probably the strangest of the Continental pepperboxes is that known to collectors as the Apache Knuckle-duster. This combination of firearm, dagger, and brass "knucks" is supposed to have been popular with Parisian gangs noted for their violence. One of these pepperboxes is pictured, Figure 171, with dagger extended and ready for use as a firearm, and also, Figure 172, with dagger and grip folded. The mechanism is double-action, with the mainspring outside the frame. The gun is marked L. DOLNE INVUR.

A much rarer type of this misbegotten weapon is illustrated in Figure 173. On this the dagger only is folding.

On both these piece the frames and "knucks" are of brass. Some were made for rim-fire cartridges, but these two both use pin-fire.

An elaborate, finely made, and gold inlaid fist pistol is pictured in Figure 174. This was specially made for Jerome Napoleon Bonaparte, grandnephew of Napoleon First, by Devisme of Paris. The initials of Colonel Bonaparte are inlaid in gold on the right of the frame. The gun uses a peculiar centerfire tapered cartridge and may be opened for loading when a gate over the recoil-shield is pushed up. This gate can be raised and the cylinder tilted up only after the hammer is drawn back slightly. An ejector rod is screwed in the butt.

A twentieth-century double-action German pepperbox is shown in Figure 175. This is called the Deadless Pistol. Presumably the name derives from its alleged ability to render a victim unconscious without fatal results. The cartridge is an early form of gas shell. It produces a great flash due to the lycopodium in it. Its irrespirable fumes, if carefully inhaled, may produce asphyxia. If discharged in a man's face, the damage could be severe, but it is nearly worthless as a weapon of defense. One can imagine a practical joker thinking it a great invention, but no one else.

Chapter 10
?? PEPPERBOXES ??

UP TO NOW the firearms pictured and described have all conformed to the definition in the first chapter. All had barrels that encircled a central axis; all fired only one barrel at a time; all had only one striker. Students and collectors will agree all were of pepperbox construction, though some examples—such as My Friend and Apache Knuckleduster—are usually referred to by their special names and not thought of as being pepperboxes.

The first form of firearm to be called a pepperbox was a gun that "peppered" or "sprinkled" with several shots fired at the same time, just as a table pepperbox sprinkles pepper through many perforations simultaneously.

Mr. Charles Winthrop Sawyer was a man of knowledge and experience. His writings, particularly his series of books, "Firearms in American History," have been of great service to students of obsolete American guns. Mr. Sawyer defined a pepperbox as *"a hand firearm having three or more barrels all to be fired by the same striker."*

It is to be noticed that Mr. Sawyer did not rule out as pepperboxes certain types that do not conform to the definition in Chapter 1. Hand firearms with three or more barrels in line, and those which fire all barrels at once, were not excluded in his concept.

Most of these aberrant forms of pepperboxes—quasi-pepperboxes, if you prefer—are considered in this chapter.

Brief consideration will be given to certain pieces which are often called pepperboxes—but which satisfy neither Mr. Sawyer's definition nor my arbitrary delimitation—and also to transitional pieces and conversions.

One of the most interesting of these ? ? Pepperboxes ? ? is illustrated in Figure 176. This is called a Duckfoot Pistol, though it is sometimes known as a Mob Pistol, and is reputed to have been popular with prison guards and tough sea captains. The four barrels all fire at once, the bullets going in different directions within a sector of a circle. The weapon was a strong deterrent to mobs or gangs of holdup men in close quarters. A man in a mob was given pause by the knowledge that if he attacked, he jeopardized his fellows. This flint' piece was made by Twigg of London. It has a solid brass frame into which the barrels are screwed, and it is supplied with a belt hook. One occasionally finds a single-shot flint pistol that has been converted to resemble a Duckfoot.

Another rare multishot weapon that fires all barrels at once is pictured in figure 177. This was made by J. Hunt, London, about the same time as the Twigg Duckfoot, 1770-1780. It has four screw barrels and was designed to place four bullets simultaneously in a small group, rather than to spread them.

Still another firearm that discharges all barrels simultaneously, but primarily intended for neither offense nor defense, is shown in Figure 178. The United States Patent Office classifies it as an Animal-Trap. It is termed a Game-Shooter by the inventors, Henry S. North and John O. Couch, in their specifications in Patent #24,573, granted June 28, 1859. This ugly device is said to have had some popularity in Australia for trapping kangaroos. In operation, a short cord fastened to the eye at the muzzle had bait attached to its loose end. One end of another cord was tied to the eye on the backstrap, and the other end to a tree or stake, leaving the suspended pistol free to move when the cord attached to the muzzle was disturbed. When the bait was taken, the cord to which the bait was fastened was pulled taut, and the barrels lined up with the cord. The tightening of the cord released the circular hammer which was driven by a heavy spring against the single cap. The six barrels discharged together

Fig. 176 Twigg—9" overall—four-shot—.50 caliber.

Fig. 177 J. Hunt—10" overall—four-shot—.38 caliber.
(JOSEPH KINDIG, JR., COLLECTION.)

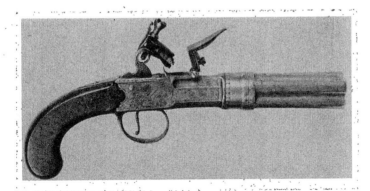

with the head of the unlucky animal directly in the line of fire. These trap pistols have the six barrels concentric with the cylinder axis, but the barrels may be drilled either parallel to the axis or at a slight angle. The fire from the percussion cap goes directly to only one chamber. The fire from the first explosion passes around an annular channel to fire the other barrels. The intervals between explosions are so short that the shots are considered simultaneous. When the gun is cocked by pulling the large round hammer fully back, it may be fired by pulling either the trigger or the rod protruding from the muzzle.

A scarcer type of the North & Couch trap pistol is shown in Figure 179. This is similar to the patented type in that it may be secured and fired by a pull on the rod running through the cylinder. The conspicuous difference is the elimination of the heavy round hammer and the substitution of a thumb hammer.

The six-barrel mob pistol illustrated in Figure 180 had all barrels fired from a single fall of the flint, but the explosions were successive. They probably made one brief continuous roar with no apparent intervals between shots. Each barrel, after the first, was fired by ignition from the preceding discharge. This Roman candle effect was managed by having the bores decrease in length, with each succeeding load and vent closer to the muzzle. The first barrel, opposite the pan, has an inside length of $7\frac{1}{4}$ inches. The inside lengths decrease progressively, down to a final $5\frac{7}{8}$ inches.

Quasi-pepperboxes having three or more barrels in line usually have the barrels one above another. An interesting example of such a gun with flint ignition is shown in Figure 181. This gold inlaid piece has the tap action mechanism that was widely used in flint pistols with two superposed barrels and sometimes in pistols with three or four barrels that encircled a central axis. (Examples of such pepperboxes were described in Chapter 2.) This very flat English pocket piece made by Durs Egg has three vents, all leading down, opened by pressing down the tap action lever on the left of the frame.

On May 26, 1857, W. W. Marston of New York obtained U.S. Patent #17386, claiming a mechanism designed ". . . to

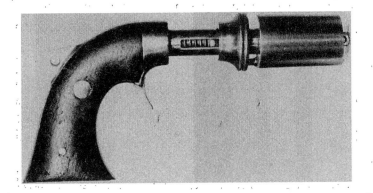

Fig. 178 North & Couch—7¼" overall—six-shot—.28 caliber.

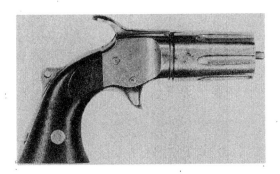

Fig. 179 North & Couch—6" overall—six-shot—.31 caliber.
(WILLIAM M. LOCKE COLLECTION.)

explode successively the barrels arranged vertically over each other . . ." Mr. Marston specified the invention was applicable to firearms with "two, three, or more' barrels placed one above the other," and contemplated its use on percussion cap weapons. A thin steel cylinder—controlled by a ratchet—that oscillated on trunnions when the hammer was cocked was set vertically in the frame. An "exploder," or striker, was carried through the flat cylinder. This moved upward to strike nipples in succession when the hammer was cocked.

It is thought that Mr. Marston did not put the percussion cap pistols in production. He did, however, have some success with cartridge pistols, using the patented mechanism with slight modification. One of these single-action pistols that follows the usual pattern, except that it has pearl grips and elaborate engraving, is shown in Figure 182. It is marked WM. W. MARSTON PATENTED MAY 26, 1857 NEW YORK CITY IMPROVED 1864. In the patent design, the oscillating cylinder was set for the initial discharge by release of a catch under the frame. In the gun illustrated a projection of the turning piece protrudes though the right of the frame. When being loaded, the turning piece is turned manually clockwise until the firing pin reaches the lowest point. The position of the projection serves to indicate which barrels are unfired. Release of a simple catch permits the barrels to be tipped down for loading. A three-pronged extractor is provided.

In advertising, Mr. Marston mentioned that his pistol "is more effective, the ball receiving the full force of the charge, there being no loss of power by gas escaping between the joints, as in all Cylinder Pistols." He also noted that the gun could be loaded in the dark "as it is unnecessary to detach any part for that purpose." He could have added what collectors sadly realize—that removable parts get lost.

A rarer model of the three-barrel Marston is pictured in Figure 183. This has a 3-inch retractable dagger. It has no extractor and lacks in the marking "Improved 1864." This early model uses .22 rim-fire cartridges instead of the .32 rim-fire used in later models.

Figure 184 shows a double-action German Reform Pistole

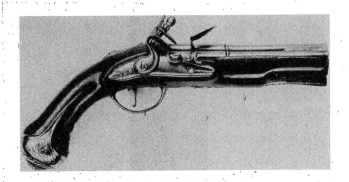

Fig. 180 Unmarked—14" overall—six-shot.
(WILLIAM M. LOCKE COLLECTION.)

Fig. 181 D. Egg—5" overall—three-shot—.34 caliber.
(GLODE M. REQUA COLLECTION.)

made for modern .25 ACP cartridges. The four barrels are in a vertical block. After the top barrel is fired, another pull on the trigger lifts the block and fires the cartridge in the second barrel. As the second barrel is fired, gas escapes from it through a small hole into the first barrel and ejects the shell from that barrel. The empty case from barrel two is ejected when barrel three is fired, and barrel three is cleared when the lowest barrel is fired. It is necessary to remove the case from the lowest barrel with a rod when the barrel block is removed for reloading. This repeating pistol, like the next to be mentioned, has the objectionable feature of a part that must be detached for loading. The Reform has a safety lock on the left of the frame. The gun is extremely thin and flat and has been a popular weapon on the Continent to carry in evening clothes.

Mr. J. Jarre, of Paris, France, obtained U. S. Patent #35685 on June 24, 1862, for a double-action repeating cartridge gun with a horizontally-sliding cartridge container. Mr. Jarre apparently at first considered making his unique firearm with the sliding member serving only as a series of chambers, all shots passing through a single barrel. Guns of that construction are more commonly found than those in which the sliding piece comprises multiple barrels, as does the example illustrated in Figure 185. These guns are usually called Harmonica Pistols because of the resemblance of the sliding breech bar to a harmonica. Trigger action moves the group of barrels horizontally and fires the cartridges successively. When a catch is released, the entire barrel group may be removed and the cartridges loaded or unloaded after a face-plate is lifted. This face-plate prevents cartridges dropping out of chambers when the loaded barrel block is secured in the frame. An ejector rod is provided, carried screwed in the butt. In this sort of gun there is no limit to the number of chambers that may be provided, but ten was considered the practical maximum.

All the firearms mentioned in this chapter up to this point have complied with the conditions of Mr. Sawyer's definition of a pepperbox.

The descriptions to follow will be of transitional pieces and

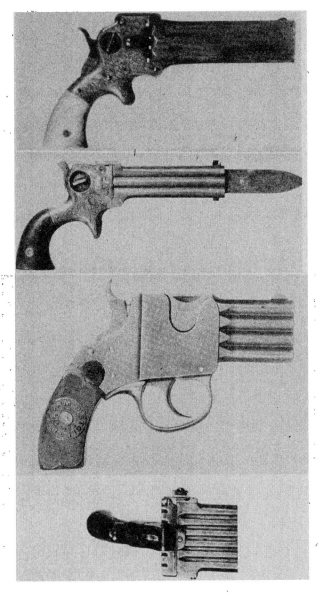

Fig. 182 (*top to bottom*) Wm. W. Marston—7¼" overall—three-shot—.32 caliber—serial #18.

Fig. 183 Wm. W. Marston—8¼" overall—three-shot—.22 caliber—serial #1091. (ROBERT ABELS COLLECTION.)

Fig. 184 Reform—5½" overall—four-shot—.25 caliber.

Fig. 185 J. Jarre—5½" overall—six-shot—serial #309.

of certain unusual pieces which are sometimes called pepperboxes but which transgress all accepted definitions.

Percussion-cap pepperboxes evolved from percussion cap "turn-over" pistols, which, when having four or six barrels, have themselves been classed by some collectors as pepperboxes. The four-barrel "turn-over" pistols are in effect double double-barrel pistols, fitted with both right- and left-hand single-action locks. Usually the two lower barrels may be brought under the hammers by simply twisting the barrel-group, just as hand-revolved cylinders on pepperboxes are turned. A spring catch permits turning in only one direction, and it clicks when the barrels are lightly secured in firing position.

Figure 186 shows an English four-barrel pistol of this type. It is marked CHs JONES, 32, Cockspur Street, LONDON. The frame and dolphin hammers are finely engraved, and the hammers have half-cock safeties. A rammer is carried in the butt. On this pocket piece the barrels are drilled in a solid block.

Another four-barrel pistol of considerable interest is shown in Figure 187. This more powerful weapon has the name ELSNER silver-inlaid in each of the barrel ribs, but no other identifying marks. The consensus of opinion is that the gun is Austrian. To turn over the barrel group, it is necessary to release a plunger by pressing back the forward end of the spring-steel trigger guard. There is a large metal-lined trap for bullets in the butt. The snap cover for this large trap itself contains a shallow trap for caps. The oddity of this double trap is enhanced by its having a coursing dog engraved on the outer cover. Unlike the Jones pistol, this has four barrels brazed together. The illustration shows the covers of both traps open.

A very unusual "turn-over" pistol with six barrels is pictured in Figure 188. This has two locks, and the barrels are turned like the barrels of the Jones pistol, but there are three pairs of barrels instead of two pairs. Each nipple has its vent in line with a barrel axis, but the nipples enter the barrels at a 45° angle. Thereby they accomplish the same purpose as the peculiarly slanting nipples used on the Irish pepperbox described in Chapter 8, having a vent cutting the powder chamber in an

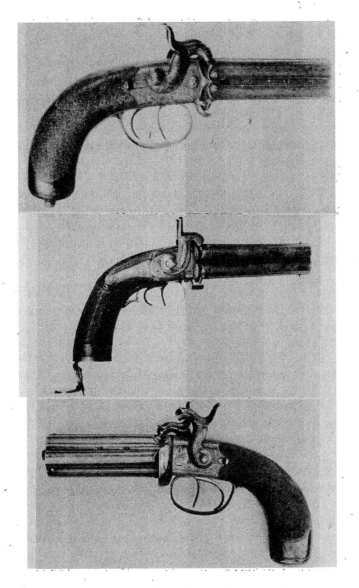

Fig. 186 (*top*) Charles Jones—8″ overall—four-shot—.38 caliber.

Fig. 187 (*center*) Elsner—9½″ overall—four-shot—.46 caliber.

Fig. 188 (*bottom*) J. Purdey—7½″ overall—six-shot—.38 caliber—serial #21. (W. KEITH NEAL COLLECTION.)

ellipse and increasing the area of the opening to the chamber. With this Purdey construction it was not necessary to bend the hammer. In this piece a rammer is carried in an opening drilled in the center of the one-piece cylinder. Barrel ribs with sights are placed between pairs of barrels and on each of the three ribs is the marking PURDEY LONDON with the London Gunmakers Proof. The gun being by Purdey, the workmanship is of course of first quality.

The three English "turn-over" pistols that have been described all have both front and rear sights. They are all fully developed and finely made weapons. They cannot be fired as rapidly as the later double-action pepperboxes, but they lack the disturbingly heavy trigger pull of those guns and can be fired more accurately.

Any praise given the transitional pieces preceding the percussion cap pepperboxes cannot be carried on to the transitional pieces that followed and evolved from the pepperboxes.

Figure 189 shows an English revolver of the transition period. It is nothing but a pepperbox with a shortened cylinder and a barrel screwed on the shortened and threaded center pin. This first form of pepperbox conversion had the serious defect of a barrel fastened at one point only. In consequence, as a result of either mishandling or just moderate wear, the barrel was apt to turn out of line.

Later pieces were constructed so as to keep the barrel bores in line with chambers, the common practice being to attach to a projection at the rear of the barrel a supporting strap running beneath the cylinder to the frame. This was a considerable improvement, but these transitional pieces were nearly all double-action and they retained the inherent disadvantages of the double-action pepperboxes—heavy trigger pull and a tendency to explode several barrels at once, there being no partitions between the nipples.

The bottom straps on these later "pepperbox-revolvers" were commonly fastened with cap screws. The illustrated example, Figure 190, has the strap held by a wing screw. The hammer has a projection under the nose, so it may rest between nipples,

Fig. 189 Unmarked—12″ overall—six-shot—.40 caliber.

Fig. 190 Charles Rosson—11½″ overall—six-shot—.36 caliber.

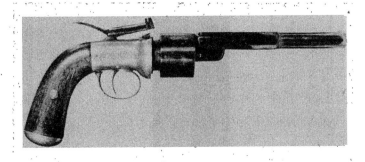

Fig. 191 Baker—11½″ overall—six-shot—.42 caliber.

and also a rectangular opening, cut through it near the fulcrum, intended to serve as a rear sight. The marking on the barrel of this British gun is CHAS. ROSSON MAKER.

The unusual transitional revolver shown in Figure 191 has the barrel held at two points by screws running into the center pin. Like all the transitional pieces, it has a mechanically revolved cylinder, but this is quite unusual in that it is single-action, with a hammer reminiscent of the Stocking pepperbox. The hammer is cut away and curved, so that a notch cut in it serves as a rear sight. There is a sliding-bar safety on the left of the frame which locks the hammer in the position shown in the illustration. The hammer is marked BAKER'S PATENT. The patent relates to the safety only. Actually, Mr. Whitmore Baker simply left at the Office of the Commissioner of Patents in London a provisional specification, which was nothing more than a petition asking that he be protected on an invention for an "Improvement in Fire-Arms Adapted to Prevent their Accidental Discharge," but he never got around to supplying drawings and the specification.

An early French piece that may conceivably antedate even the regular form of pepperbox is shown in Figure 192. This has a form of ignition in use before percussion caps were invented. Detonating material placed in the vents leading to the chambers was ignited by the blow of the projection on the hammer. If the barrel be removed, what remains is a single-action pepperbox with a mechanically-revolved cylinder. As a guard against multiple explosions, spring covers fit over the two vents adjacent to the vent under the hammer. The marking is blurred, but appears to be MICHALLON BREVETE.

The percussion cap gun illustrated in Figure 193, made by Josiah Ells of Pittsburgh, Pennsylvania, is sometimes called the Ells Pepperbox-Revolver. Again we have pepperbox construction with a barrel screwed on the cylinder pin. This is double-action with a mechanically revolved cylinder. The hammer is stamped ELLS PATENT on the left and the patent dates AUG. 1, 1854 and APRIL 28, 1857 on the right. The feature of the first patent—a lock capable of use either single- or double-action—

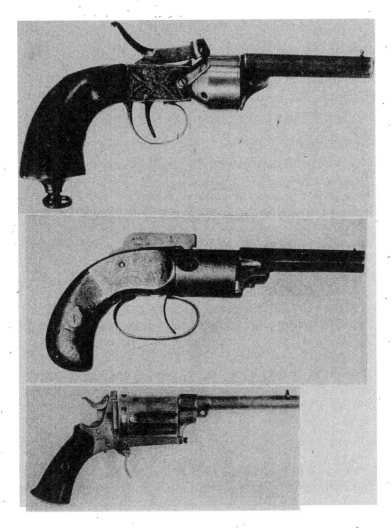

Fig. 192 (*top*) Michallon Brevete—8″ overall—six-shot—.32 caliber.
Fig. 193 (*center*) Ells—7¼″ overall—five-shot—.28 caliber.
Fig. 194 (*bottom*) Unmarked—7½″ overall—six-shot.
(MORTENSON COLLECTION—photograph courtesy *The Gun Collector*.)

was not embodied in the gun illustrated. The gun has the second invention, a mechanism for revolving the cylinder by use of a lever moving in radial grooves in the rear face of the cylinder. The nipples are set at a 45° angle like the nipples in the Purdey six-shot "turn-over" pistol. One slot is cut in the cylinder between two nipples so the hammer may rest without touching a capped nipple.

Even the Continental fist pistols were converted to revolvers. That the revolver shown in Figure 194 was not originally manufactured as a revolver is evidenced by the fact that the barrel is smooth bore, while the chambers are rifled. The gun was completed, without the addition of a barrel, as a pepperbox using center-fire cartridges. The barrel was probably added when the maker found no market remained for fist pistols.

The two Reid revolvers in Figures 195 and 196 are pepperbox conversions only by extension of the term. The inducement to Mr. Reid to produce these outgrowths of My Friend is mentioned in Chapter 7. In principle they are knuckle-dusters with barrels added, but in the construction of the frames they are distinct variations. Both are rare, but the one with the fluted cylinder is particularly so. This has a steel frame and is marked REID'S NEW MODEL .32 MY FRIEND. The other has a brass frame.

One of the startling creations of Herman of Liege, some of whose pepperboxes were described in Chapter 8, is shown in Figures 197 and 198. Of all the makers of pepperboxes, Herman was surely the least repressed. This remarkable double-action piece has four fixed barrels welded together and four separate hammers operated by one folding trigger. Successive trigger pulls raise and drop the hammers one after the other from right to left and fire the barrels in counter-clockwise order. The gun is equipped with front and rear sights and a rammer screwed into the center of the cylinder. It bears Belgian proof marks and is marked HERMAN BREVETE.

Probably the most extraordinary percussion cap gun that has been photographed for this book is the Harrington pistol shown in Figures 199 and 200. It has seven barrels, two hammers, and two triggers—but it has only one nipple and can be fired only

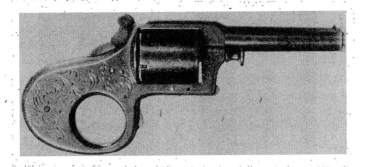

Fig. 195 James Reid—7" overall—five-shot—.32 caliber.

Fig. 196—James Reid—6" overall—five-shot—.32 caliber.
(GLODE M. REQUA COLLECTION.)

once without reloading. All barrels are discharged together when the right-hand hammer falls. As the patent papers are available, most of the construction is understandable. The patent papers make no mention of a second lock. The effect of the installation of the second lock on this piece will be explained, but the exact reason for the installation will remain uncertain.

Henry Harrington, of Southbridge, Massachusetts, describing himself as a cutler, obtained United States Patent #297 on July 29, 1837. The one claim in his patent was for "the throwing of shot or balls from any number of barrels united together, by exploding powder in a single chamber . . ." The patent referred to a long gun with a sliding, removable powder chamber, and to a screw-barrel pistol without such a chamber. The actual pistol illustrated here has both a screw barrel and a sliding powder chamber. What appears to be a single barrel is a cylinder drilled with seven barrels, six encircling one in the center.

Mr. Harrington's specification stated: "In the breech or lower end of the barrel is a mortise . . . to receive a powder chamber. This chamber is made nearly square, is exactly fitted to the space made by the mortise, and slides therein . . . The front plate of this chamber has holes drilled through it to receive the shot or balls . . . The chamber itself contains a cavity sufficiently large to hold a proper charge of powder. It is covered with a flat plate of metal, turning on a pin at one corner."

One illustration is a side view with the right hammer cocked. The other illustration shows the mortise cut in the "lower end of the barrel," and the chamber with its cover opened.

The means of fastening the chamber in the mortise is explained by Mr. Harrington: "Behind this chamber a screw is placed, turning into a thick piece of metal corresponding in position to the breech-pin of the common gun. The head of the screw is large enough to admit of receiving a small handle or pin, which projects under the barrel . . . (By moving the pin sideways) the screw is made to press hard against the chamber, and to hold it firmly in its place, or, by turning in the opposite direction, to leave the chamber so loose that it can slide out easily."

Fig. 197 Herman—9½" overall—four-shot—.48 caliber.
(FRANK R. HORNER COLLECTION.)

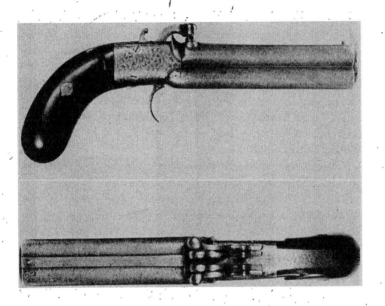

Fig. 198 Another view of Fig. 197.

Though not mentioned in the patent, the "handle or pin" is fitted so it may be set in a new position when inevitable wear requires increased tightening.

The procedure necessary to get the gun ready to shoot is: "The shot or balls to be used are poured or placed in the holes on the outside of the front plate of the chamber, and pressed into them by the finger. The cover is then turned open and the powder poured into the chamber . . . The cover is again closed, the chamber slipped down into the mortise of the barrel, the percussion-cap placed on the tube, the screw turned up so that the head presses the chamber hard against the lower ends of the small barrels, and the gun is then ready to be discharged."

The barrel casing is of silver. The butt cap is of ivory. The rear sight has an open leaf screwed on, so a slight adjustment for elevation is possible.

The gun may be fired by raising only the right hammer and pressing the forward trigger. If both hammers be raised, the gun can be fired by pressing the forward trigger only after the left hammer has been released by pulling the rear trigger. Therefore the left hammer serves as a safety. This very unusual safety appears to have been an afterthought. The engraving on the gun follows the outline of the right hammer but not of the left hammer. The lock plate is cut back for the right hammer in order to stop the hammer fall in case the hammer be snapped with no chamber in position, but the left plate is not cut back. The trigger plate appears to have been made for a single trigger and then recut for two triggers.

The appending of the second hammer was clearly not planned when Mr. Harrington started to build this gun. The sole result of the change of mind is a safety feature, but it seems possible Mr. Harrington may have considered using two hammers to supply dual ignition. This could have been accomplished by the use of a chamber with two nipples. Designing the gun to explode two caps at each discharge would be a safeguard against a defective or lost cap on one nipple.

The pistol is marked on the barrel with the name HARRINGTON and with the word CUTLER, and also with HENRY HARRINGTON

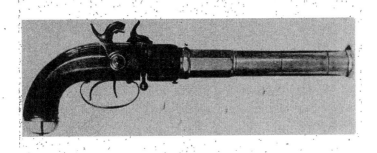

Fig. 199 Harrington—12½″ overall—seven-shot—.24 caliber.
(FRANK R. HORNER COLLECTION.)

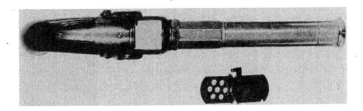

Fig. 200 Another view of Fig. 199.

PATENT—SOUTHBRIDGE, MASS. As no patent date appears, it could not have been made later than 1842.

Probably the most extraordinary cartridge gun that has been photographed for this book is pictured in Figures 201 and 202. This well-made and finely finished Italian piece is remarkable because of its multiplicity of unusual features that are not found in other guns. It has two separate groups of barrels, and in each group the barrels diverge both vertically and horizontally. Eight charges may be exploded either all at once or in two bursts of four. Two hammers each strike two firing pins, and each pin strikes two cartridge heads. The trigger, a flat button under the barrel, is moved forward when a hammer is cocked and is fired by a squeeze of a finger tip.

In Figure 201 the horizontal spreading of the barrels at the muzzle may be seen, and in Figure 202 the vertical spreading. A burst of shots spreads both sideways and upward and downward. When a catch over the recoil-shield is released, the pivoted group of eight barrels may be tipped down for loading. There is a unique clip that folds over the cartridge heads, permitting the insertion of eight shells in the barrels in one movement. There are four vertical slots cut in the back of the clip, openings for the firing pins. Each pin strikes one top cartridge and also the cartridge directly underneath. When the left hammer only is cocked, pressure on the trigger fires the left group of four cartridges, the hammer acting on two vertical firing pins. When both hammers are cocked, trigger pressure fires all eight shots simultaneously. In addition to having unusual features, the little gun has style. The marking on the barrels is MITRAGLIERA PRIVILEGIATA SISTEMA MEROLLA GIOVANNI.

Fig. 201 Italian double pepperbox—5½" overall—eight-shot—.21 caliber.
(GOVERNOR GORDON PERSONS COLLECTION.)

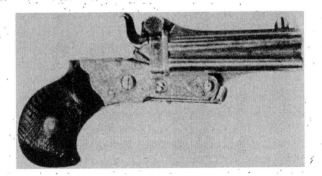

Fig. 202 Another view of Fig. 201.

Chapter 11
PEPPERBOX SHOULDER PIECES

LONG-BARREL pepperboxes fired from the shoulder were made in comparatively small numbers. They are much rarer than one-hand pepperboxes.

Fitting just two long barrels in satisfactory alignment is difficult. Getting several barrels in alignment is virtually impossible.

The long pepperboxes were costly to produce. They were also heavy and clumsy. As barrels increase in number, the gun's weight becomes objectionable.

Most of the long flint pepperboxes were made by the celebrated English maker, Henry Nock. They exist both with revolving barrels from which separate shots were fired, and with fixed barrels from which all shots were fired in a single burst. These latter, commonly called volley guns, are of two types, sporting and military.

Figure 203 shows one of the extremely rare revolving flintlock coaching carbines. Because of the ingenious magazine primer, one required only a little practice to fire six shots with some rapidity. Cocking the hammer, closing the pan cover, and giving the barrel a one-sixth turn between shots took less time than it takes to explain the operation.

Henry Nock produced this remarkable piece. It has six smooth-bore barrels brazed around a tube which houses the shaft connected to the frame. The hand-turned barrels turn clockwise and are indexed by a ratchet wheel and detent.

The feature that makes the gun of great interest is the automatic priming magazine. A supply of priming powder is contained in the hollow frizzen. A snap cover in the top of this

Fig. 203 Henry Nock—33½" overall—six-shot—.48 caliber.
(DR. PRENTISS D. CHENEY COLLECTION.)

Fig. 204 Henry Nock—36½" overall—seven-shot—.50 caliber.
(DR. PRENTISS D. CHENEY COLLECTION.)

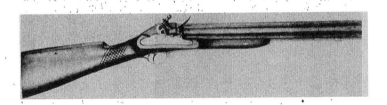

Fig. 205 Henry Nock—37" overall—seven-shot—.50 caliber.

priming powder magazine permits filling, and a revolving door in the base regulates the emptying of powder into the flash pan. This revolving door is a cup which holds powder for one filling of the pan. It is linked to the front of the lock plate and turns, when the frizzen is moved, within the limits imposed by the linkage. When the blow of the hammer raises the frizzen, the mouth of the cup turns into the magazine and powder fills the cup. When the frizzen is lowered, the cup turns back until its mouth is over the pan and powder is dropped in the pan. Most of the later Collier revolving flint guns used the same form of priming magazine, though the first Colliers, at least one of which is believed to have been made by Samuel Nock, employed a magazine with an outside ratchet and no linkage.

The entire gun shows the fine craftsmanship of "best" guns. Fitting of parts is exact. The vents are gold-lined and the panel, engraved H. NOCK, set in the lock plate, is of gold.

The Nock volley guns were first made for sportsmen. They had seven barrels—six encircling one in the center—that were all discharged at once. As sporting firearms, with each barrel charged with two or three large cast slugs, these guns would reach out and bring down geese at long range.

Guns of this type, with shorter barrels and larger bores, were made by Henry Nock a few years later for use in combing out the fighting tops of enemy ships. As a result of that use, guns like the two illustrated in Figures 204 and 205 are often called Crow's Nest or Lord Nelson guns. These seven-barrel carbines were adopted by the British Royal Service after Lord Nelson was killed at Trafalgar by a shot from a mizzen top.

The first of these pieces—once the most prized long gun in the famous collection of the late Albert Foster, Jr.—has the usual flint lock and is without the heavy forestock of the second Lord Nelson gun. The second gun has a Bolton lock. These excellent locks, patented by George Bolton in 1795, were all made by Henry Nock. The Bolton lock contains no screws. The parts are made to guage and fitted between two plates with the sear and tumbler working on pins fixed to the plates. Though these were probably the first locks precision-made for military use,

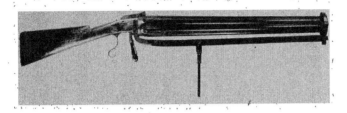

Fig. 206 J. R. Cooper—49″ overall—six-shot—.60 caliber.
(DR. PRENTISS D. CHENEY COLLECTION.)

Fig. 207 Pieper—42″ overall—seven-shot—.22 caliber.
(DR. PRENTISS D. CHENEY COLLECTION.)

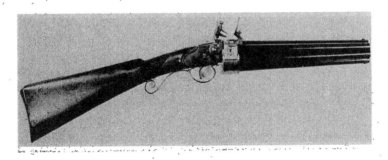

Fig. 208 Unmarked—30″ overall—seven-shot—.50 caliber.
(SMITHSONIAN INSTITUTION COLLECTION.)

with interchangeable parts, and though they were bought in quantity by the British Government, they were fitted to few guns. J. N. George states that despite the large purchases of these exceptional locks, British muskets or carbines which have been equipped with Bolton locks are the rarest Service arms of their day. The top rib of the barrel is marked H. NOCK LONDON GUN-MAKER TO HIS MAJESTY and the lock is marked H. NOCK.

On the Continent some sporting flintlock guns were made with fourteen barrels and two locks. Two groups of seven barrels each were placed side by side and fired as conventional two-barrel shotguns.

The gun shown in Figure 206 is a top hammer percussion cap pepperbox long gun and, because of its unusual features, a very interesting example. It is marked J. R. COOPER, MAKER and has Birmingham proof marks. The maker did not use his patented concealed hammer mechanism on this piece. Perhaps this gun antedates his patent. It is a shoulder gun fitted for use as a swivel gun. There is no conventional trigger. Instead, there is a handle which when pushed forward revolves the barrels, and which when pulled back fires the hammer. Most of these long multi-barrel guns, both flint and percussion, were smooth bore. This has six deeply rifled barrels and may have been intended for use on small boats carrying landing parties.

Henry Nock's idea of a many-barreled gun firing all barrels at once remained popular with some hunters for wildfowl shooting through the percussion cap era and was even continued in cartridge guns. Figure 207 is of a "goose gun" made by Pieper of Liege, using the Remington rolling-block action and firing seven .22 rim-fire cartridges simultaneously. What appears in the illustration as a single heavy barrel is a tube containing seven rifled barrels. The seven cartridges are held in a breech plate separate from the barrels; they are all fired at once by a multiple firing pin and are all ejected by one operation.

A flintlock shoulder piece that operates similarly to the seven-barrel pepperbox described in Chapter 2 is shown in Figure 208. This has a single lock and frizzen, and seven bronze barrels turned by hand.

Chapter 12
CASED SETS AND ACCESSORIES

SOME OF THE pepperboxes in the cased sets illustrated in Figures 209 to 224 have already been described in previous chapters. Others, being quite like pepperboxes already described, will require no descriptions here.

Cases for pepperboxes were not made with the ornamental inlaid tops sometimes found on cases for pairs of pistols, nor were they supplied with elaborate accessories. They were often of mahogany or oak, sometimes of rosewood or ebony. A case for a percussion pepperbox contained a mold and a flask as essentials. A loading rod, a screwdriver, a nipple wrench, a box of caps, an oil can, and a nipple pricker were other accessories, some of which were usually included also.

A case for a cartridge pepperbox often held nothing but the gun and space for cartridges. Sometimes there was a space for a box of shells; sometimes there were holes for individual shells.

In instances where a gun requires a special accessory, that special tool may be a very scarce item. An example is the tool for unscrewing Robbins & Lawrence barrels, shown in the illustration of the cased Robbins & Lawrence.

An array of various types of flasks supplied with cased American pepperboxes is shown in Figure 225. The four bag-shaped flasks are types which are also found with revolvers; the other flasks are distinctive to pepperboxes.

Accessories from one English cased set are shown in Figure 226.

The last illustration, Figure 227, is of a holster for a pepperbox. It is of the army type, with a flap and a loop through which the belt passes. The closed end is made large to accommodate a pepperbox muzzle.

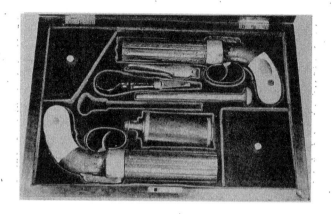

Fig. 209 Presentation pair of Allen pepperboxes. Ivory grips. Elaborate engraving on silvered nipple shield and on blued frame and barrels. Silver plate engraved: "To Norton McGiffin, Co. K, 1st Reg. Pa. Vol. Presented by the citizens of Washington, Pa. for his gallantry at the Siege of Puebla, Oct. 12, 1847". (DR. PRENTISS D. CHENEY COLLECTION.)

Fig. 210 Robbins & Lawrence cased pepperbox.
(GEORGE N. HYATT COLLECTION.)

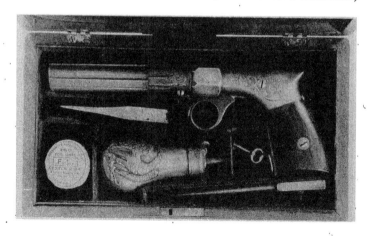

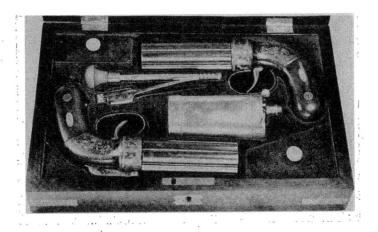

Fig. 211 Cased pair of J. G. Bolen pepperboxes, made by Allen.
(DR. PRENTISS D. CHENEY COLLECTION.)

Fig. 212 Henry Tatham of London cased pepperbox.
(DR. PRENTISS D. CHENEY COLLECTION.)

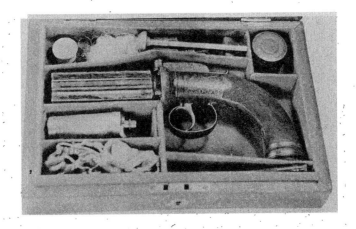

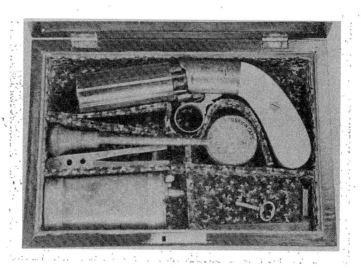

Fig. 213 Blunt & Syms cased American pepperbox.
(DR. PRENTISS D. CHENEY COLLECTION.)

Fig. 214 Charles Jones of London cased pepperbox.
(DR. PRENTISS D. CHENEY COLLECTION.)

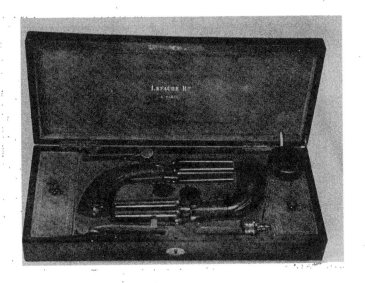

Fig. 215 Le Faure of Paris cased pair of Mariette pepperboxes.
(DR. PRENTISS D. CHENEY COLLECTION.)

Fig. 216 Chaumont of Liege cased Mariette pepperbox.

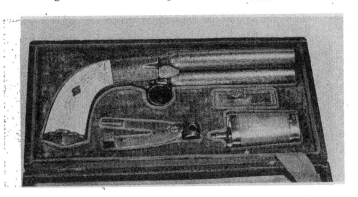

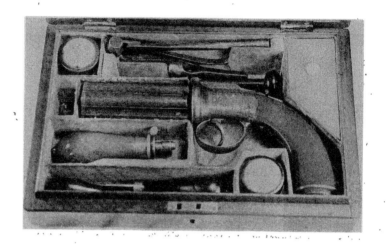

Fig. 217 T. Boss of London cased pepperbox.
(DR. PRENTISS D. CHENEY COLLECTION.)

Fig. 218 Allport of Cork cased pepperbox.
(DR. PRENTISS D. CHENEY COLLECTION.)

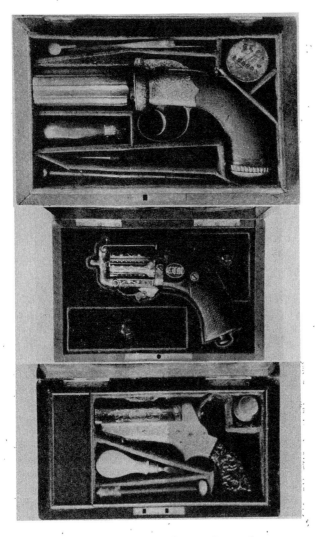

Fig. 219 (*top*) Baker of London cased pepperbox.
Fig. 220 (*center*) Devisme of Paris cased pepperbox.
Fig. 221 (*bottom*) Tipping & Lawden cased pepperbox.

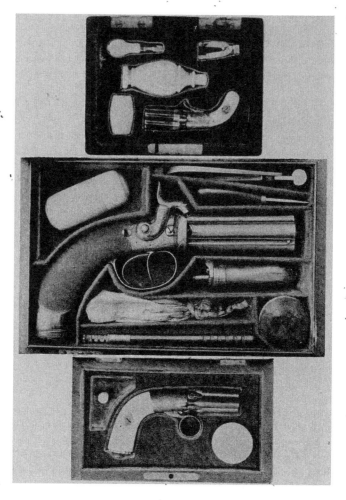

Fig. 222 (*top*) Miniature cased pepperbox. (DR. W. R. FUNDERBURG)

Fig. 223 (*center*) Purdey of London cased pepperbox—or six-barrel "turn-over" pistol. (W. KEITH NEAL COLLECTION.)

Fig. 224 (*bottom*) Belgian cased Mariette pepperbox.

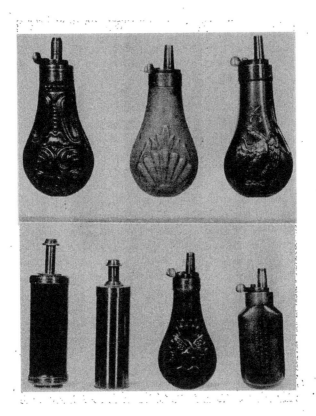

Fig. 225 Powder flasks used with American cased pepperboxes.
(JOSEPH L. SPICER COLLECTION.)

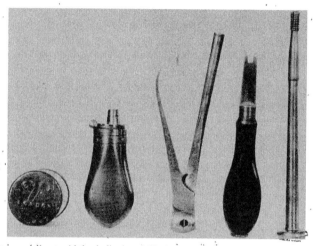

Fig. 226 Accessories supplied with a British cased pepperbox.

Fig. 227 Holster for pepperbox.
(HAROLD G. YOUNG COLLECTION)

BIBLIOGRAPHY

Abridgements of the Specifications Relating to Fire-Arms and Other Weapons, Ammunition and Accoutrements. London, 1859.

Antiques. (Periodical) Various Issues.

Army and Navy Journal. (Periodical) Various Issues.

BLANCH, H. J. *A Century of Guns.* London, 1909.

BOCK, GERHARD. *Moderne Faustfeuerwaffen und Ihr Gebrauch.* Neudamm. First edition, 1911.

COLT, SAMUEL. *Sam Colt's Own Record.* Hartford, 1949.

DEANE, J. *Deane's Manual of the History and Science of Fire-Arms.* London, 1858. Facsimile reproduction, 1946.

Dictionary of American Biography. New York, 1928 et seq.

GARDNER, ROBERT EDWARD. *Arms Fabricators Ancient and Modern.* Columbus, Ohio, 1936.

GARDNER, ROBERT EDWARD. *American Arms and Arms Makers.* Columbus, Ohio, 1938.

GEORGE, JOHN NIGEL. *English Pistols and Revolvers.* Onslow County, North Carolina, 1938.

GEORGE, JOHN NIGEL. *English Guns and Rifles.* Plantersville, South Carolina, 1947.

GREENER, WILLIAM WELLINGTON. *The Gun and Its Development.* London, 1881.

THE GUN REPORT. (Periodical) 1939 to 1942. Various Issues.

HARRISON, G. CHARTER, JR. *The Gun Collector.* (Periodical) Currently published at Whitewater, Wisconsin. Formerly published as *The Gun Collector* and *The Gun Collectors Letter* at Madison, Wisconsin. Various Issues.

KARR, CHARLES LEE, JR. & KARR, CARROLL ROBBINS. *Remington Handguns.* Harrisburg, Pennsylvania, 1947.

KITCHENER, MAJOR, H. E. C. "Revolvers and Their Use." *Journal of the Royal United Services Institutions,* Vol. 30, No. 136, London, 1886.

Frank Leslie's Illustrated Newspaper. (Periodical) Various Issues.

Magazine of Antique Firearms. (Periodical) Athens, Tennessee, 1911 to 1912. Various Issues.

METSCHL, JOHN. *The Rudolph J. Nunnemacher Collection of Projectile Arms.* Milwaukee, 1928.

National Cyclopedia of American Biography. New York, 1898 et seq.

PARSONS, JOHN E. *The Peacemaker and Its Rivals.* New York, 1950.

POLLARD, CAPTAIN HUGH BERTIE CAMPBELL. *The Book of the Pistol and Revolver.* London, 1917.

POLLARD, MAJOR HUGH BERTIE CAMPBELL. *A History of Firearms.* London, 1926.

The Rifle. (Periodical) Succeeded by *Shooting and Fishing*, later by *Arms and the Man*, and finally by *The American Rifleman*. 1885 to date. Various Issues.

SATTERLEE, L. D. *A Catalog of Firearms for the Collector.* Detroit, 1927.

SATTERLEE, L. D. *Ten Old Gun Catalogues for the Collector.* Detroit, 1940.

SATTERLEE, L. D. & GLUCKMAN, MAJOR ARCADI. *American Gun Makers.* Buffalo, 1940.

SAWYER, CHARLES WINTHROP. *Firearms in American History.* Boston, 1910.

SAWYER, CHARLES WINTHROP. *Firearms in American History. Vol. II. The Revolver; 1800 TO 1911.* Boston, 1911.

Scientific American. (Periodical) Various Issues.

Stock and Steel. (Periodical) Marshalltown, Iowa, March to July, 1923. Various Issues.

MARK TWAIN. *Roughing it.*

Turf, Field and Farm. (Periodical) Issue October 16, 1874.

VAN RENSSELAER, STEPHEN. *American Firearms.* Watkins Glen, New York, 1947.

& BRAY, Sole Agents,
No. 245 Broadway, N. Y.

in the Army and Navy.
VOLVERS!!
Best Cartridge Revolver yet invented, 6in. Barrel, 80

Broadway, New York.
Send for a Circular.

Subscription Books for Canvassers Agents.

G. P. PUTNAM'S Publishing A
532 Broadway, New York.
...sers, Book Agents, Postmasters, &c.
...lowing Works, publishing in NUMB
...ive Cents each, EXCLUSIVELY for Ca
...genta, will be found well worthy your atten
That a large sale of these Works can be secur
active and energetic canvass in every City, Town,
lage in the country, there is no doubt. Every h
person desires to preserve the story of this great
the National Life. The
REBELLION RECORD
is the work best adapted to give to every one the
plete and reliable History to be found. The
HEROES AND MARTYRS
will contain faithful and correct Likenesses, eleg
graved on steel, of the *Generals, Military* and N
roes, *Statesmen, Orators,* and the *Leaders of th*
tion, with succinct and reliable Biographies of ea
will form, when completed (which is expected b
Two handsome Quarto Volumes of Twenty Num
each), a work of permanent interest and value.
Agents are wanted in every City, Town, and Vi
the United States, with whom liberal arrangements
made, to canvass their respective localities.
All applications for Agencies must state if they
present engaged in canvassing for any work; if so
Apply, by letter or otherwise, to
CHAS. T. EVANS, Gen. Agent, 532 Broa
, Specimen Numbers of either will be sent on
of 12c. in Postage Stamps.

---

or Cheap
ELETS.
ETS.
CHAINS.
CHAINS.
INDS OF SETS.
INDS OF PINS.
TURE PINS OF ALL
o for full particulars.
URING JEWELER.

ES.
s. Enclose stamp for
, 208 Broadway, N.Y.

— Hegeman & Co.'s
Paint, Grease Spots,
ks, &c., equal to new
color or fabric, only
sts. Be sure and get
AN & CO., N. Y.

good number of
month, or will allow
ory. Business of the
irees, with stamp en-
x 2552, Boston, Mass.
and gentlemen wish-
es without retarding

Wanted in every
to a respectable and
rs address, with red
H. WARNER,
Street, New York.

ODIST
vorite Songs, which,
dollars, will be sent,
ty-five cents, by
ON & CO., Boston.

ENTS ARE
's PAT HEMMER and
new articles of ready
nt free on receipt of
np for price-list and
way, N. Y.

icted with *Rheumat-
old* and *Frost-bitten*
y wearing *Metlam &
llic Insoles*, will find
their use, never hav-
Office 429 Broadway.
ON, 2013 Girard Ave-

n Printer.
31 Park Row N. Y.

for this disease is the
end postage stamp for
Y.

OST MONEY.
y selling our " PRIZE
ACKAGES." Circu-
Beekman St., N. Y.

vous Debility,
ko.

Life Pills.
y composed from the
tea districts of China,
arned Chinese physi-
of eating opium, &c.

DATE DUE

53

ELLIOT'S POCKET REVOLVER.

A MOST POWERFUL ARM, WHICH CAN BE carried constantly about the person without inconvenience or danger. Length four inches, scarcely more than that of the barrels. It is the most compact, safe, and powerful Pocket Revolver ever made; weighs only eight ounces, charged with cartridges, which can be purchased in hardware stores, each barrel rifled, gain twist and sighted. Will penetrate one inch of pine at one hundred and fifty yards. Send for illustrated Circular.
Retail price, Plated Frame, with 100 cartridges, $10 00
"   " Blued Frame,   "   "   "   9 50
Trade supplied.   T. W. MOORE, 426 B'way, N. Y.

WEDDINGS supplied with the new style of Marriage Cards and Envelopes, by A. DEMAREST, Engraver, 182 Broadway. Seals and stamps.

Commercial Travelers and Agents Wanted to Sell our 25 Cent Portfolio Package.
Contents.—18 Sheets Note Paper, 18 Envelopes, 1 Penholder, 1 Pen, 1 Pencil, 1 Blotting Pad, 100 Recipes, 1 War Hymn, 5 Engravings, 1 New Method for Computing Interest.—2 Fashionable Embroidery Designs for Collars, 4 for Under-Sleeves, 2 for Under-Skirts, 1 for Corner of Handkerchief, 2 for Cuffs, 1 for Silk Purse, 1 for Child's Sack, 1 for Ornamental Pillow Case, 1 Puzzle Garden, and ONE BEAUTIFUL ARTICLE OF JEWELRY. $10 a day can be realized. Send stamp for Circular of wholesale prices.
WEIR & CO., 43 South Third Street, Phila., Pa.

$75 A MONTH!—I WANT TO HIRE AGENTS in every County at $75 per month and expenses, to sell a new and cheap Sewing Machine. Address (with stamp)   S. MADISON, Alfred, Maine.

A New Edition of
Harper's War Map
Now Ready.

Price Six Cents.

Sent by mail on receipt of price.
HARPER & BROTHERS, New York.

The New Issue of Postage Stamps, of all denominations, for sale. Apply to
HARPER & BROTHERS, Franklin Square, N. Y.

Wedding Cards and Note Papers at J. EVERDELL'S celebrated Engraving Establishment, 302 Broadway, cor. Duane Street, N. Y. Samples by mail.

SOMETHING NEW — Agents w
to make $50 to $100 a month, selling our
patented articles, wanted in every family. Selling
ly. Satisfaction guaranteed. Samples 25c. each,
stamp.   RICE & CO., No. 53 Nassau Street,

HARPER'S
NEW MONTHLY MAGAZI
TERMS.

One Copy for one Year . . . . . $
Two Copies for One Year . . . .
Three or more Copies for One Year (each) .
*And an Extra Copy, gratis, for every Club of
SUBSCRIBERS.*

HARPER'S MAGAZINE and HARPER'S WEEKLY, to
one year, $4 00.
Clergymen and Teachers supplied at the
CLUB RATES.
The DEMAND NOTES of the United States will be r
for Subscriptions. Our distant friends are request
mit them in preference to Bank Notes.
HARPER & BROTHERS, PUBLISHERS,
FRANKLIN SQUARE, NEW Y

HARPER'S WEEKL
Single Copies Six Cents.

A Thrilling Story,
Entitled,

NO NAM
By Wilkie Collin
AUTHOR OF
"The Woman in White."
Richly Illustrated by John McLe
Was commenced in the Number for March 15 (No.
HARPER'S WEEKLY,
And will be continued from week to week until co

TERMS.
One Copy for One Year . . . . .

www.ingramcontent.com/pod-product-compliance
Lightning Source LLC
LaVergne TN
LVHW010338170225
803910LV00002B/22